# Elementary Mathematical Structure

Merrill Mathematics Series

Erwin Kleinfeld, Editor

*George R. Vick*
Sam Houston State College

# Elementary Mathematical Structure

**Charles E. Merrill Publishing Co.**
*A Bell & Howell Company*
*Columbus, Ohio*

Copyright © 1969 by Charles E. Merrill Publishing Co., Columbus, Ohio. All rights reserved. No part of this book may be reproduced in any form, electronic or mechanical, including photocopy, recording, or any information storage and retrieval system without permission in writing from the publisher.

Standard Book Number 675-09574-3

Library of Congress Catalog Card Number: 69-10631

1 2 3 4 5 6 – 74 73 72 71 70 69

Printed in the United States of America

*Preface*

There seems to be a great diversity of opinion concerning the type of undergraduate-level course best suited to preparing students to teach mathematics in the elementary grades. Some books are concerned primarily with the acquisition of arithmetic competence through exhaustive practice; others introduce the prospective teacher to as much modern mathematics as possible which has any bearing upon the topics of elementary school mathematics. In my opinion, textbooks of both types must be liberally supplemented by the instructor—the first type, with material to develop mathematical insight, and the other, with suggestions and illustrations pointing up the relation of the more abstract and sophisticated textbook material to its grade-level counterpart.

This book attempts to create a reasonable compromise of the often conflicting demands of (1) college students who may be relatively poorly prepared in mathematics, (2) the elementary school child's limited scope of mathematical experience, (3) the need to develop skill in both teacher and child, and (4) the college instructor who is alert to the need for logical development. This effort sometimes makes the treatment seem inconsistent. Most of that which appears to be inconsistent is the result of deliberate choice.

Although teachers need not become mathematicians in order to teach mathematics acceptably in a modern elementary program, they do need to know how mathematicians *think*. For this reason, I have included material on logic and the structures of number systems.

Since the initial need for the idea of *sets* in the primary grades is to establish *number* concepts and the concept of *addition*, introductory material on sets is presented first without any reference to numbers, even as elements of sets. Generally, the treatment of the theory of *cardinal numbers* ignores the *ordinal* property. The introduction of addition of whole numbers in the primary grades, however, relies heavily on the elementary *counting* process, which necessitates the early introduction of the ordinal property of the *natural* numbers. Very little written communication about numbers is possible without some knowledge of our *numeration* system. The use of multi-digit numerals is purposely avoided, however, until *bases* and *place value* are discussed in Chapter 3.

The mathematical importance of the principle of *induction* and the attendant intuitive appeal of the Peano concept of addition (essentially by repeated addition of ones) are discussed in Chapter 4. The remainder of the chapter can be omitted, since the properties developed there are also developed from the cardinal number concept in the following chapters. In fact, most of Chapter 4 is suitable only for the stronger students.

I feel strongly that *geometric concepts* are perceptually different from arithmetic or *numerical* concepts, and that *distance* is a concept already better-developed in the mind of the first-grade child than any other basic geometric concept. Since it is doubtful that anyone can appreciate the distinctive properties of non-euclidean geometry before he is acquainted with at least a few of the basic euclidean properties, it seems more reasonable to capitalize on this intuition than to ignore it. Furthermore, since young children are less expert at freehand drawing than adults, I am in favor of lending them a straight edge and a drawing compass so that their number line drawings will *look* like number lines.

The number line is introduced with the *whole* numbers rather than with the *real* numbers for three reasons: (1) it is used in the elementary grades at a corresponding level of development; (2) early introduction allows time for familiarization; and (3) the *incompleteness* encourages interest in and speculation about the existence of other types of numbers. For similar reasons, cartesian plane coordinates and measure of rectangular area are included with the development of the *cartesian product* of sets and multiplication of whole numbers.

The concepts of *subtraction* and *division* as operations inverse to addition and multiplication may never disappear. They also reinforce the concept of *inverse elements*. Since equation-solving has become a fundamental concept of the elementary mathematics curriculum, and since it has been significant in the development of the idea of *extensions* of algebraic systems, this technique is used to develop the *integers* and the *rational numbers*. There are three reasons for introducing the *integers* ahead of the *rational* numbers: (1) this order allows natural usage of standard names for the new numbers; (2) it follows the order of introduction of the corresponding operations; and (3) students have less trouble with integers than with rational numbers. Although it is a topic which belongs to the system of whole numbers and integers, the concept of *prime number* is introduced in the study of common fractions, since it provides one of the earliest and best applications in elementary arithmetic.

Although it would be futile to attempt a logically coherent and rigorous development of the system of real numbers at this level, I believe the elementary teacher should be given as much insight as possible into a working knowledge. It seems to me that Cantor's approach to these concepts is much better suited to this purpose than either the Dedekind treatment or the abstract algebraic description of a complete ordered field.

To avoid breaking the continuity of the theme of expanding number systems, other important topics are postponed until the final chapter. *Percentage* is given a very brief treatment, because I have found that the technique of writing and solving equations is a meaningful and efficient way to handle the subject. *Approximations* are handled as intuitively as possible, with heavy reliance upon definition. The object is not the development of skill or theorems, but of the basic understanding which enables one to establish most accurate bounds for the number being calculated from approximations. In the discussion of *probability* I have taken some liberty with terminology in an attempt to use descriptive terms.

You will find in the exercise sections that problems marked with an asterisk involve particularly important ideas or valuable supplementary material.

I should like to express my gratitude to the mathematicians and mathematics educators who have reviewed this material and offered helpful suggestions for its improvement. Special thanks go to my colleague, Professor Dan Reeves, who collaborated with me in class-testing the material. Finally, I must express my deep appreciation to the editors of Charles E. Merrill Publishing Co. for their patience and encouragement during the entire process of writing, revising and publishing.

*George R. Vick*

# To the Student

This book contains the most basic and elementary material in mathematics, and the treatment is generally modern. There are a few places, however, where the approach is somewhat old-fashioned. This occurs in cases where I consider the conventional approach superior to the modern from the pedagogical viewpoint, or where it is not feasible to use the modern approach.

You should avoid adopting either one of two extreme attitudes concerning this material. One attitude is the fear which arises from previous unhappy experiences with mathematics or from concern about complete lack of contact with the "new math." Actually, one of the most important reasons for the modern approach to learning mathematics is that it *helps* the student to understand. If you will study for understanding and will seek the assistance of your instructor when you do not understand this text, you will realize that mathematics is not a mystery. The other dangerous attitude to avoid is that of overconfidence, which can lead to lack of study. The material in this course may have been at your fingertips at one time, but it is highly unlikely that it still is.

I have tried to make the material interesting, but not by artificial means. In other words, if you find this course interesting, it will be because you find mathematics interesting (perhaps for the first time). Your instructor and I both hope you do.

*George R. Vick*

# Table of Contents

| | |
|---|---|
| **LANGUAGE AND MATHEMATICS** | 1 |
| *Chapter 1* <br> **SETS** | 15 |
| *Chapter 2* <br> **CARDINAL AND ORDINAL PROPERTIES** | 25 |
| *Chapter 3* <br> **NUMERALS AND BASES** | 37 |
| *Chapter 4* <br> **THE ORDINAL PROPERTY OF $W$** | 53 |
| *Chapter 5* <br> **ADDITION OF WHOLE NUMBERS** | 65 |
| *Chapter 6* <br> **MULTIPLICATION OF WHOLE NUMBERS** | 89 |

*Chapter 7*
**INVERSE OPERATIONS**     **109**

*Chapter 8*
**EXTENDING NUMBER SYSTEMS**     **129**

*Chapter 9*
**EXTENDING THE DECIMAL NUMERATION SYSTEM**     **163**

*Chapter 10*
**SPECIAL TECHNIQUES AND APPLICATIONS**     **185**

**SELECTED ANSWERS**     **221**

**BIBLIOGRAPHY**

**INDEX**

# List of Symbols

| | | | |
|---|---|---|---|
| $\rightarrow$ | Implies | $J$ | Set $\{\cdots {}^-2, {}^-1, 0, 1, \cdots\}$ of integers |
| $\leftrightarrow$ | If and only if | | |
| $\sim$ | Is not (Logic) | $F$ | Set of rational numbers |
| $\vee$ | Or | $R$ | Set of real numbers |
| $\wedge$ | And | $n'$ | Successor of whole number $n$ |
| $\in$ | Is an element of | | |
| $\notin$ | Is not an element of | ${}^-n$ | Negative of number $n$ |
| $S$ | Set $S$ | $\bar{n}$ | Reciprocal of number $n$ |
| $\{x:$ statement about $x\}$ | Set-builder notation | $(S; *)$ | Number system involving set $S$ and operation $*$ on its elements |
| $\subseteq$ | Is a subset of | $(S_1, S_2, \cdots, S_n)$ | Sequence (ordered list) of $n$ objects |
| $\subset$ | Is a proper subset of | | |
| $\sim$ | Is equivalent to (sets) | $\|ab\|$ | Distance between points $a$ and $b$ |
| $\not\sim$ | Is not equivalent to | | |
| $\{\,\}, \varnothing, N_0$ | The null set | $a(b)c$ | Point $b$ is between points $a$ and $c$ |
| $U$ | Universal set | | |
| $S$ | Complement of set $S$ | $\overline{ab}$ | Line segment having end points $a$ and $b$ |
| $\cup$ | Union | | |
| $\cap$ | Intersection | $\overrightarrow{ab}$ | Ray having end point $a$ |
| $\times$ | Cartesian product (sets) | $\overset{\frown}{abc}$ | Angle with vertex $b$ |
| $=$ | Equals, or is equal to | $\%$ | Per cent |
| $\neq$ | Is not equal to | $\approx$ | Approximates or is approximated by |
| $>$ | Greater than | | |
| $<$ | Less than | $P[E]$ | Probability that event $E$ occurs |
| $\geq$ | Greater than or equal to | | |
| $n(S)$ | Cardinal number of set $S$ | $m(S)$ | Measure of a set $S$ |
| $N$ | Set $\{1, 2, 3, \cdots\}$ of natural numbers | $n!$ | $n$ factorial or $n(n-1)(n-2)\cdots 2 \cdot 1$ |
| $W$ | Set $\{0, 1, 2, \cdots\}$ of whole numbers | $\binom{n}{k}$ | $\dfrac{n!}{k!(n-k)!}$ |
| $N_n$ | Ordered set $\{1, 2, 3, \cdots, n\}$ | | |

# Language and Mathematics

Language is a means of communication. In some situations the need for language is slight, perhaps because those who are in communication are engaged in very simple activities where few ideas are involved, as in the daily lives of primitive tribes; or perhaps everyone involved is so familiar with the details that each can perform his part with few instructions, as in a surgical operation, for example.

Certainly the process of obtaining a formal education at the college or university level does not fall into this category. This complex procedure involves the communication of hundreds of ideas daily, many of them new to the student, and some quite abstract or complicated. Sometimes these ideas are represented by words and symbols which are new to the student. More often, though, they are described by familiar terms used in a slightly different sense, or in an unfamiliar combination with other known terms. In any event, it is essential that the student become well-acquainted with the language of the subject which he hopes to master. This familiarity does not mean simply the ability to spell the words and give their definitions. It means, rather, that the student understands the full meaning of the entire statement in which the words were used and that, in turn, he can use the language correctly to convey to someone else his own thoughts on the subject.

Although extensive use is made of special symbols in mathematics, the basic language by which it is described in this country is English. Thus, the reader is encouraged to make full use of his knowledge of the English language, rather than waste his time speculating as to what unexpected mathematical meaning may be borne by some perfectly common English word!

## THE USE OF SPECIAL SYMBOLS

Since the meanings of words are determined by the people who use them, we find these meanings changing from generation to generation and from place to place. So, it is often efficient to invent a new symbol for a concept. An extensive set of such symbols has been devised in *symbolic logic*. Although these symbols are translated into words as they are read, the unique meanings of the symbols are accepted while the words with their various connotations are ignored. Most modern mathematicians use such symbols, especially when they are anxious to express their ideas succinctly and precisely.

To illustrate the occasional inadequacy of language, the following classic example is presented:

>A dog has one tail more than no dog.
>No dog has eight tails.
>Thus, a dog has nine tails.

Of course the difficulty here lies in the multiple meaning and usage of the word *no*. Another example occurs in a statement which frequently appears in mathematics textbooks:

>"... all *objects* which belong to both *Ann* and *Bill*."

(The words in italics have been substituted for certain technical terms, but the sense of the statement has not been affected.) The question is, does the collection of objects consist of all objects belonging to Ann together with all objects belonging to Bill? Or, does it consist of all objects each of which is jointly owned by Ann and Bill?

This last example shows that even mathematicians, who are among the most exasperatingly precise people on earth, are not always as precise as they would like to be. Although we will make every effort to be as lucid as possible, it is unlikely that you will interpret correctly everything in this book. It is important, therefore, that you take advantage of every opportunity to discuss these ideas with your instructor and with your classmates.

The remainder of this chapter is concerned with a brief study of certain aspects of *logic*, or *foundations*. The object of the study is to examine the formal nature of mathematical thinking and some of the technical meanings of certain standard statements made in English.

## STATEMENTS

Although mathematics asks many questions, as do other areas of study, its focus is on the answers or the factual information which it derives. In our study of grammar we usually designate a collection of words which makes such an assertion of fact a declarative sentence. Here we shall use the shorter term *statement*, although some people still prefer to use the word *sentence* to mean the same thing.

We find in our daily experience that statements may agree with facts or they may not. If a statement does agree with fact we say it is *true*; if it does not agree, it is said to be *false*. We will not delve into the philosophical question, "What is truth?," but rather confine our discussion to commonly accepted "facts." (Speculation about the accuracy of our interpretation of the mathematical properties which we perceive through our five senses belongs to a subject called Metamathematics.)

Does every declarative sentence fall exclusively into one of the categories *true* or *false*? In the investigation of this question, the following example has been concocted:

"This statement is false."

Suppose the example is indeed a statement. If it is true, then it must also be false, since it *says* it is false. On the other hand, if it is false, it must also be true, since it agrees that it is false. Before drawing conclusions, let us examine another illustration of a declarative sentence. Suppose someone says, "I am thinking of two numbers whose sum is ten." Is the statement true or false? Until the two numbers are made known, the classification of the statements is unknown. If the sentence had been

$$3 + 4 = 10$$

or

$$2 + 8 = 10$$

then it would have been a *statement*. However, such a sentence, often written in the form

$$x + y = 10$$

is not a statement because its "truth value" is indeterminate. Such sentences are important to the study of mathematics, however, and some people call them *open sentences*.

From observations such as we have made, the Greek philosopher Aristotle formulated three basic laws of logic:

1. *Law of Identity* (a thing is itself).
2. *Law of the Excluded Middle* (a statement is either true or false).
3. *Law of Non-Contradiction* (a statement is not both true and false).

The Law of Identity insures against changing the rules in the middle of the game, so to speak. When a symbol is chosen to represent an object, it must represent that same object whenever it occurs in the context of that particular discussion. The use of the expression "either ⋯ or" in the explanation of the Law of the Excluded Middle illustrates the meaning of this phrase as used by mathematicians. If "either true or false" had meant "true or false, but not both" then there would have been no need for the Law of Non-Contradiction. Thus when we use the word "or," or the phrase "either ⋯ or," we shall allow the possibility that both alternatives hold true simultaneously. When we wish to rule out the case of both alternatives holding true, we shall simply add the word *exclusively*, as in "either ⋯ or ⋯, exclusively."

To summarize and formalize the essential elements of the foregoing discussion we now present a formal definition. Such formality will be observed from time to time in order to provide an easy reference to the technical terms needed for future communication. The student should be aware, however, that the formal definition is seldom sufficient to produce understanding—for this it is necessary to study the context.

*Definition 1.* A statement is a declarative sentence which conforms to Aristotle's laws of logic.

## Exercise 1

1. Does the existence of a language imply the existence of intelligence? Support your answer with an argument based upon the first sentence of this chapter.
2. Is there much correlation between the degree of intelligence of members of a culture and the degree of sophistication of their language? You may wish to confer with language and cultural specialists before attempting to resolve this question.
3. In what sense is the word "or" used in the following statement? "Bill is either a sophomore or a junior."
4. Discuss the use of the word "or" in the statement, "Either I have gained weight or the cleaners have shrunk my trousers."
5. In a certain village it is said that the village barber shaves everyone who does not shave himself. If the barber is a man, who shaves him?
6. Discuss the possible meanings of the sentence, "Harold caught a fly."
7. Write a sentence which has more than one meaning.
8. Try to find one of the classic comedy routines which rely on the multiple usage of words for their humor. Examples: "Who's on first?" and "Why is a fire truck red?"
9. Is the sentence, "Roses are red" a statement? Explain your decision.
10. What kind of sentence is "This rose is red" as relates to truth or falsehood?

## CONNECTIVES

One of the attributes of the rational mind, essential to every kind of human progress, is its ability to associate ideas. In fact, association of ideas is such a common activity of the human mind that it is the basis of many tests given by psychologists in their attempts to detect mental abnormality.

In mathematical thinking, just a few types of association of ideas are sufficient to express even the most complex relationships. The three connectives most commonly employed by mathematicians are expressed by the words *and, or* and *implies.* As grammatical parts of speech, the first two of

## Language and Mathematics

these three words are "coordinating conjunctions," and are used to connect two ideas of equal importance. In the language of formal logic, the words *and* and *implies* are called "conjunction" and "implication," respectively, while the word *or* is called a "disjunction," probably because it seems to imply the choosing of one of two alternatives to the exclusion of the other. As we previously remarked, however, the use of the word in mathematical logic does not exclude the acceptance of both alternatives.

Distinctions in the meaning and usage of these connectives can best be understood by means of a "truth value" analysis. A compact and systematic method for such an analysis is a so-called *truth table*. The truth table is compact because it uses simple symbols to represent words and statements. The following example illustrates the use of symbols, connectives and a truth table.

EXAMPLE: Suppose a student is absent from his first period class and the professor says, "Either John overslept or he is ill." Is the professor's statement true or is it false?

In the analysis of this statement we shall let the letter $p$ represent the statement "John overslept," and the letter $q$ shall represent the statement "He is ill." If we use the standard symbol "$\vee$" to represent the word "or" or the concept "either $\cdots$ or $\cdots$," the composite symbol

$$p \vee q$$

represents the statement

"Either John overslept or he is ill."

In the truth table below we show all the possible truth values of this statement, based upon the possible truth values of the simple statements $p$ and $q$.

*Table 1.*

| Statements | $p$ | $q$ | $p \vee q$ |
|---|---|---|---|
| | T | F | T |
| Truth | F | T | T |
| Values | T | T | T |
| | F | F | F |

The first row of truth values in the table represents the case in which John overslept but was not ill, so the professor's statement was true, as indicated in the last column. The second row shows the case where John did not oversleep but he was ill, so the professor's statement would be true in this case also. In the third row we see that John overslept and he was ill also, and the professor's statement is still true. But in the last case, John did not oversleep nor was he ill; he got up early and went fishing! In this case the professor's statement was not true.

This example illustrates the fact that of all possible combinations of truth and falsity for any two statements $p$ and $q$, *the only case in which the statement $p \vee q$ is false is where both $p$ and $q$ are false.*

## NEGATION

You may have noticed in the discussion of the preceding example that the word "not" occurred a number of times in a rather natural way, primarily to point out that a particular statement was false. Since it is often needed in discussions about statements, a special symbol "$\sim$" has been adopted by many logicians to represent the word "not" or the idea of *negation*. Thus the statement $\sim p$ (read "not $p$") says, "The statement $p$ is false." You should understand that $p$ and $\sim p$ both represent statements either of which may be true, but not both, because of the Law of Non-Contradiction. Thus, the symbol $p$ is equivalent to the assertion

"The statement $p$ is true"

while the symbol $\sim p$ says

"The statement $p$ is false"

or

"The statement $\sim p$ is true"

The following truth table should help to clarify the foregoing discussion. Although $p$ represents any statement, to be more specific you may let $p$ represent the statement "John overslept," so that $\sim p$ would represent the statement "John did not oversleep."

*Table 2.*

| $p$ | $\sim p$ | $\sim(\sim p)$ |
|---|---|---|
| T | F | T |
| F | T | F |

The observation that a statement $p$ and its negation $\sim p$ cannot both be true is an example of the use of the connective *and*. Before examining that particular statement, however, we shall construct a truth table for an example using more familiar statements. The symbol "$\wedge$" will be used for the word *and*.

EXAMPLE: A true-false statement on a history exam said, "Lafayette came to America and fought for the British." Although everyone correctly marked the question *false*, some noticed that the statement, "Lafayette came to America" is true, while the statement "He fought for the British" is false. If $p$ and $q$ respectively denote these two statements, the first row of truth values in Table 3 show the true situation, while other possibilities for the statements $p$ and $q$ are shown in the rest of the table.

**Table 3.**

| $p$ | $q$ | $p \land q$ |
|---|---|---|
| T | F | F |
| F | T | F |
| F | F | F |
| T | T | T |

This table reflects the principle that *for a statement of the form $p \land q$ to be true, it is necessary that both $p$ and $q$ be true.*

From the preceding example and observation, when $p$ is any statement, in order for $p \land \sim p$ to be true, both $p$ and $\sim p$ must be true. But if $\sim p$ is true, this means that $p$ is false, so that $p \land \sim p$ would mean that $p$ is true and $p$ is false, which is not possible because of the Law of Non-Contradiction. Therefore, if $p \land \sim p$ is to be classified as a statement, we must conclude that it is a false statement; that is,

$$\sim(p \land \sim p)$$

is a true statement.

## Exercise 2

1. Use a truth table to analyze the statement, "Either you prepare your lessons or you fail the course."

2. If $p$ represents the statement, "Men prefer blondes" and $q$ the statement, "Ann is a blonde," write in words the following statements:

    (a) $\sim p$    (b) $p \lor \sim q$    (c) $p \land q$    (d) $\sim(p \land q)$

3. If $p$ represents the statement, "It is two o'clock" and $q$ the statement "The bus is late," use $p$ and $q$ to write symbols representing the following statements:
    (a) It is two o'clock and the bus is late.
    (b) Either the bus is late or it is not two o'clock.
    (c) The bus is not late.
    (d) It is two o'clock but the bus is not late.

4. From your own knowledge classify each of the following statements as true or false:
   (a) It is not true that George Washington was the first president of the United States of America.
   (b) This statement is neither true nor false.
   (c) A truth table determines whether a statement is true or false.
   (d) I (the reader) marked the statement in part (a) true.

In the following problems let $p$ and $q$ represent unknown statements.

*5. Compare truth values of $p \vee q$ and $q \vee p$.

*6. Compare truth values of $p \wedge q$ and $q \wedge p$.

*7. Use a truth table to compare the three statements $\sim(p \vee q)$, $\sim p \vee \sim q$ and $\sim p \wedge \sim q$.

*8. From the results of Problem 7, find a statement which has the same truth values as $\sim(p \vee \sim q)$.

*9. Under what truth values of $p$ and $q$ is $p \wedge q$ false?

*10. Use a truth table to compare $\sim p \vee \sim q$ with $\sim(p \wedge q)$.

## IMPLICATION AND LOGICAL EQUIVALENCE

One of the simplest ways in which two statements can be related is by *logical equivalence*. If $p$ and $q$ are (logically) equivalent we write $p \leftrightarrow q$ and mean that $p$ and $q$ have the same truth value. From the truth table involving negations we see that $p$ and $\sim(\sim p)$ are equivalent. Thus, for every statement $p$ we have the following:

$$p \leftrightarrow \sim(\sim p)$$

It might be interesting here to restate Definition 1 in terms of the new symbols and concepts which we have been discussing. Compare the following statement with the earlier definition to see if they have the same meaning to you.

*Definition 2.* A *statement $p$* is a declarative sentence for which (1) $p \leftrightarrow p$, (2) $p \vee \sim p$, and (3) $\sim(p \wedge \sim p)$.

Of all the ways in which statements can be related, the one most useful to mathematics is probably *implication*. We say the two statements "John is older than Mary" and "Mary is older than Jim" imply the statement "John is older than Jim." By this we mean that if the first two statements are true then the third statement is true also. Let $p$, $q$ and $r$ represent these statements in the order listed; then

$$p \wedge q \rightarrow r$$

represents the statement "If John is older than Mary and Mary is older than

Jim, then John is older than Jim." The statement $p \wedge q$ introduced by the word "if" in our word version is usually called the *hypothesis* by mathematicians, while the statement $r$ which follows the word "then" is called the *conclusion*. Logicians often prefer to call these parts the *antecedent* and the *consequent*, respectively; these terms imply "that which goes before" and "that which follows as a result." If these ideas are construed in their logical sense they are helpful, but their use in a chronological sense is misleading, since the "if" clause often appears in the latter part of the statement. For example, the statement "John is older than Jim, if John is older than Mary and Mary is older than Jim" is equivalent to our former statement, but in this case, the conclusion is stated first and the hypothesis follows (in the chronological sense). Notice also that the word "then" is not used in this form of the statement. Therefore if you wish to use a word clue to help you identify the parts of an implication, you should concentrate on the word "if" and the statement which it introduces.

## THE "ONLY IF" CONDITION

The reader should be warned against the dangerous practice of relying upon a single word or phrase to interpret a sentence. The same word used in different contexts can have very different effects upon the meaning. The usage of the little word "if" affords an illustration. Suppose a father tells his son, "You may see the show if you mow the lawn." Now he may have meant to say, "You may see the show *only* if you mow the lawn;" that is, he may have intended the mowing of the lawn to be a requirement or necessary condition to be met before he would allow the boy to see the show. In other words, the first statement gives the consequence of mowing the lawn while the second statement gives the consequence of not mowing the lawn. In order to examine all possibilities concerning these two statements, we shall construct a truth table for each.

Let $p$ represent the statement "You mow the lawn" and $q$ the statement "You may see the show." In reading the following table, remember that the

*Table 4.*

| $p$ | $q$ | $p \rightarrow q$ | $\sim p$ | $\sim p \vee q$ |
|---|---|---|---|---|
| T | T | T | F | T |
| T | F | F | F | F |
| F | T | T | T | T |
| F | F | T | T | T |

only case where $p \to q$ is unquestionably false is when $p$ is true and $q$ is false. In other words, the only test of the promise is to mow the lawn; then if $q$ is false the promise was false also. The fact that only one of the four cases is false reminds us that the same is true for the disjunction. Notice that the truth values of $\sim p \vee q$ are the same as those of $p \to q$. This shows that

$$(\sim p \vee q) \leftrightarrow (p \to q)$$

or $\sim p \vee q$ is equivalent to $p \to q$.

Now for the second statement "You may see the show only if you mow the lawn." From our earlier discussion, you should be able to agree that it is equivalent to the statement, "If you do not mow the lawn you may not see the show." In terms of the symbols $p$ and $q$, this statement is represented by $\sim p \to \sim q$ or $p \vee \sim q$. Its truth values are shown below:

**Table 5.**

| $p$ | $q$ | $\sim p$ | $\sim q$ | $\sim p \to \sim q$ | $p \vee \sim q$ |
|---|---|---|---|---|---|
| T | T | F | F | T | T |
| T | F | F | T | T | T |
| F | T | T | F | F | F |
| F | F | T | T | T | T |

The difference in the two statements is apparent after examining the difference in the truth tables. There are other interesting and significant comparisons involving $p \to q$, and related implications formed by using the negations of $p$ and $q$. All such implications are listed below with the equivalent disjunctive form written beneath each one. Since $p \vee q$ and $q \vee p$ are logically equivalent for all $p$ and $q$ (see Problem 5, Exercise 2) we observe that the second row of implications is a duplication of the statements in the first row.

$$
\begin{array}{llll}
p \to q & p \to \sim q & \sim p \to q & \sim p \to \sim q \\
\sim p \vee q & \sim p \vee \sim q & p \vee q & p \vee \sim q \\
\\
\sim q \to \sim p & q \to \sim p & \sim q \to p & q \to p \\
q \vee \sim p & \sim q \vee \sim p & q \vee p & \sim q \vee p
\end{array}
$$

In mathematical logic, two methods of altering an implication have proved useful: (1) reversing the order, and (2) substituting the negations of the basic statements. The application of (1) to the implication $p \to q$ produces the related implication $q \to p$ which is said to be the *converse* of $p \to q$. If (2) is applied to $p \to q$, the result is $\sim p \to \sim q$ which is called the *opposite* of $p \to q$. Referring to the entries of the right hand column in the preceding paragraph, we see that these implications are logically equivalent; that is, the converse and the opposite of an implication are logically equivalent. In symbols,

$$(\sim p \to \sim q) \leftrightarrow (q \to p)$$

A comparison of the forms of these two implications reminds us that the pair in each column of the table above are related in the same way. This observation in turn causes us to compare with $p \to q$ (from the left hand column) the equivalent $\sim q \to \sim p$ which is often called the *contrapositive* of $p \to q$. We can illustrate the word form of these related implications by applying them to our "lawn mowing" example.

EXAMPLE: If our basic implication is $p \to q$ (if you mow the lawn you may see the show), then the converse of $p \to q$ is $q \to p$ (if you saw the show, then you mowed the lawn); the opposite of $p \to q$ is $\sim p \to \sim q$ (if you do not mow the lawn you will not see the show); and the *contrapositive* of $p \to q$ is $\sim q \to \sim p$ (if you did not see the show then you did not mow the lawn). You should notice that it is often necessary to change the wording, in addition to supplying negations and changing the order of words, to express correctly the meaning of a related implication.

In studying the related forms

$p \to q$     (basic implication)
$q \to p$     (converse)
$\sim p \to \sim q$     (opposite)
$\sim q \to \sim p$     (contrapositive)

you may have noticed that the contrapositive $\sim q \to \sim p$ is the converse of $\sim p \to \sim q$ (i.e., the converse of the opposite of $p \to q$) and also the *opposite* of $q \to p$ (i.e., the opposite of the converse of $p \to q$). For this reason the contrapositive of an implication is frequently called the *opposite converse* of the implication.

## DOUBLE IMPLICATION

As we have seen, the *opposite*, and hence the *converse*, of an implication is not equivalent to the implication, as was illustrated by comparing the statements "you may see the show if you mow the lawn" and "you may see the show only if you mow the lawn." If the father had wanted to make the show both an incentive and a reward for mowing the lawn, he could have combined both implications into one statement: "You may see the show *if and only if* you mow the lawn." Such a statement is called a *double implication* and can be expressed symbolically by

$$(p \to q) \land (q \to p)$$

It is also quite customary to use such abbreviations as "iff" for the phrase "if and only if," especially when the component statements are written in words rather than symbols.

The observation that $p \leftarrow q$ seems to say the same thing as $q \rightarrow p$ might lead us to use the symbol $p \leftrightarrow q$ as an abbreviation for the double implication. However, we have already selected this symbol to signify that "$p$ and $q$ are logically equivalent;" so in order to be consistent, the only justification for using the symbol to represent the double implication also would be that $(p \rightarrow q) \wedge (q \rightarrow p)$ turns out to be logically equivalent to $p \leftrightarrow q$. The following table establishes this relationship. Indeed, the double implication is often used to define logical equivalence.

*Table 6.*

| $p$ | $q$ | $\sim p$ | $\sim q$ | $p \rightarrow q$ or $\sim p \vee q$ | $q \rightarrow p$ or $\sim q \vee p$ | $(p \rightarrow q) \wedge (q \rightarrow p)$ | $p \leftrightarrow q$ ($p$ and $q$ have same truth value) |
|---|---|---|---|---|---|---|---|
| T | T | F | F | T | T | T | T |
| T | F | F | T | F | T | F | F |
| F | T | T | F | T | F | F | F |
| F | F | T | T | T | T | T | T |

*Exercise 3*

1. If $p$ represents the statement "Black is white," and $q$ the statement, "It is two o'clock," write the following statements in words:
   (a) $q \rightarrow p$   (b) $\sim p \rightarrow q$   (c) $\sim q \rightarrow \sim p$

2. Using the symbols of Problem 1, write the following statements in symbols:
   (a) It is two o'clock if black is white.
   (b) Black is white only if it is two o'clock.
   (c) If it is not two o'clock, then black is white.

3. State the hypothesis and the conclusion of the implication in each part of Problem 2.

4. For the implication "We will be late if we do not hurry," write in words its (a) opposite, (b) converse, (c) contrapositive.

5. Let $p$ represent the statement "Black is white" and $q$ the statement "It is two o'clock." Write in words (a) $p \leftrightarrow q$ and (b) $q \leftrightarrow p$. (c) Is either of these statements more meaningful to you than the other?

6. For each meaning of $p$ and $q$ given below, discuss the truth values of $p \to q$ and $p \leftrightarrow q$.
    (a) $p$: Chris is not older than Max.
        $q$: Max is younger than Chris.
    (b) $p$: John is Mary's husband.
        $q$: Mary is John's wife.
7. An implication $p \to q$ is said to be valid iff $q$ is true whenever $p$ is true.
    (a) Write a valid implication whose converse is invalid (not valid).
    (b) Write a valid implication whose opposite is invalid.
8. "If you pass your tests and if you make at least 90% on your final, you will make a "B" in this course." With this promise you received a "C" as a grade for the course. What can you conclude?
9. If $(p \wedge q) \to r$, what does $\sim r$ imply about the truth of $p$ and $q$?
10. If $(p \vee q) \to r$, what does $\sim r$ imply about the truth of $p$ and $q$?

# 1

# Sets

## THE SET CONCEPT

Every branch of learning has certain fundamental ideas or concepts which occur in every aspect of the subject. The concept of *set* has this important relationship to mathematics. Since its usefulness is partially the result of its generality, mathematicians like to place as few restrictions as possible on this concept. A *set* is simply a collection of objects or *elements* of some kind. The objects do not have to be of the same kind, contrary to everyday usage as in "a set of dishes" or "a stamp collection." An element of a set may be either concrete or abstract. For example, such diverse objects as this sheet of paper, the first word on this page, and your feeling about mathematics, are all elements of a set.

To be of practical use in mathematics, sets must have two restrictions placed upon them. The need for one of these restrictions is illustrated by the following hypothetical case: In a certain village the men fell into two classes (sets to you!); those who shaved themselves and those who did not. Someone said, "The village barber shaves the men who do not shave themselves. Who shaves the barber?" Don't be surprised if you have trouble deciding to which set the barber belongs! To avoid such unpleasant situations, we shall insist that sets as we speak of them be well defined; that is, for each set and each object, the object is either definitely an element of the set or definitely *not* a member of the set.

The need for the second restriction can be illustrated by a list of the names of guests that a man is bringing home to dinner. While setting the table, his wife finds six names on the list, with the name "John" appearing twice. Did her husband make a mistake and list the same person twice, or is he bringing two men named John home to dinner? We shall insist that in any list of the elements of any set, each element occurs exactly once. In other words, in each set each element is *distinct* from every other element of that set.

## SET NOTATION

It is almost impossible to write about objects without using *symbols* for the objects. We have just discussed the need to distinguish one object from another. We must also be able to distinguish between different sets. Sets are commonly symbolized in three ways: (1) by a capital letter; (2) by a list of symbols (often lower case letters) enclosed in braces to represent elements of the set; and (3) by a description of the elements of the set. The following example illustrates the first two methods of symbolizing sets.

EXAMPLE: Suppose we wish to represent the set of letters of the English alphabet which we call vowels. Perhaps the most natural way to identify this set is to list the elements according to method (2). Thus,

$$\{a, e, i, o, u\}$$

represents the set which we have described. Although this notation clearly identifies the elements of the set, it has the disadvantage of being cumbersome, especially for sets containing many elements. If repeated reference is to be made to a set, it should be identified by method (1). In the case of our example, we might write the following:

$$\text{Let } V = \{a, e, i, o, u\}$$

This would mean that the symbol $V$ is to be used to represent the set whose elements are listed within the braces which follow the = symbol.

At this point, we might pause to consider the importance of definitions in mathematics. The mathematician's concept of a definition is somewhat more precise than someone else's concept and demands greater care both in its formulation and in its study. Mathematical definitions must be studied, not simply read. Their purpose is to enable the student to decide conclusively whether a particular object under study belongs to a particular classification or not. The following definition applies to every usage of the equals symbol in this book:

*Definition 1.1.* The symbol = (read equals) means that the two symbols immediately to its left and right both represent the same object.

For example, $* = \#$ means that $*$ and $\#$ are two symbols for the same object.

You are familiar with this usage of the = symbol even though you may have never seen it formally defined. In algebra you learned that $x = y$ meant that the letters $x$ and $y$ represented the same number. As we have already indicated, if $A$ and $B$ represent sets, $A = B$ means that $A$ and $B$ are symbols for the same set. You may wonder why a set should be represented by two different letters. First, $A$ and $B$ were used as two different symbols, not necessarily as letters, in the same way the symbols $*$ and $\#$ were used in Definition 1.1. You must be careful not to read into a statement in this text an interpretation from some other context in which the same symbols occur. Second, the symbols $A$ and $B$ may have been assigned when it appeared that two different sets were being represented. For example, have you ever discovered that a friend whom you know by one name is the same person known to another friend by a different name?

## SET-BUILDER NOTATION

Just as it is sometimes feasible to represent the same object by different symbols, it is helpful to allow a single symbol to represent several different objects. For example, John Doe represents each member of a large group of people. Such usage is desirable when a statement is made about an individual in a generic fashion; that is, when the statement applies equally well to many individuals.

Generic usage of a symbol occurs in the common method of representing a set by describing its elements. For example, rather than describing the elements of set $V = \{a, e, i, o, u\}$ by saying, "Each element of $V$ is an English vowel," we write:

$$V = \{* \mid * \text{ is an English vowel}\}$$

This statement is read, "$V$ is the set of all elements $*$ such that $*$ is an English vowel." The vertical bar is used merely to separate the generic element symbol from the description of the element; a colon is sometimes used instead. This expression is commonly referred to as "set-builder" notation.

EXAMPLE. The notation $\{x \mid x \text{ is a pachyderm}\}$ represents collectively all elephants in existence, while $\{y \mid y \text{ flew across the Atlantic in 1927}\}$ represents the collection of things which flew across the Atlantic Ocean in 1927, including a man named Charles Lindbergh and an airplane named The Spirit of St. Louis. The set $\{\# \mid \# \text{ is a tall letter of the English alphabet}\}$ is the set $\{b, d, f, h, k, l, t\}$.

### Exercise 1.1

1. List the elements of the set $\{* \mid * \text{ is the English name of an ocean}\}$.
2. The following lists are improper representations of sets. Write them properly.
    (a) $\{a, c, b, a, b, d\}$
    (b) $\{w \; x \; y \; z\}$ where $w, x, y$ and $z$ are elements.
    (c) $(*, \#)$
    (d) $\begin{Bmatrix} a, b, c, \\ d, e \end{Bmatrix}$

18   *Elementary Mathematical Structure*

3. Let each of the following sets be composed of members of your mathematics class as described. Show which are equal, using symbols properly.
   (a) $A$ is the set of "$A$" students.
   (b) $B$ is the set of male students.
   (c) $C = \{x \mid x$ has blue eyes$\}$.
   (d) $D = \{t \mid t$ wears trousers to class$\}$.

4. Show the set of all two letter English words having "$i$" as the first letter.

5. Represent the set of all English words composed of two letters.

6. Represent the following set another way: $\{y : y$ is a 30 day month$\}$.

7. Identify three sets which would be convenient to work with in an elementary grade classroom.

8. Describe a set you would use in teaching sets in the first grade.

*9. From the viewpoint of Definition 1.1, examine the following statements, and justify or criticize them.
   (a) $A = A$ for every $A$.
   (b) If $A = B$ then $B = A$.
   (c) If $A = B$ and $B = C$, then $A = C$.

*10. Suppose you were handed two lists of names and asked to determine whether they contained the same names. Analyze how you would do this. From this experience, write a practical definition for equality of sets.

## SETS ARE NOT NEW

You should gather from the preceding discussion that the idea of sets is neither modern nor complicated. In fact, everyone learns about numbers initially by counting the elements of some set, such as a set of fingers. As is often the case, the simple, natural concepts are the ones which can best be refined into powerful tools. If the set concept were useful only in learning to count, mathematicians would never have given it a second glance. Its value is far beyond any mathematics you can imagine from the material you will cover in this course, but you will see an astonishing number of applications even here. To make sets work for us in arithmetic, we must develop an ability to combine and compare sets with each other.

As we proceed, we will become more formal, making formal definitions, and assumptions which we will call axioms. This formality will help you focus upon the important concepts and principles, and provide you with a more precise understanding than is possible in a more casual style.

*Definition 1.2.*   $x \in A$ means an object represented by $x$ is an element of a set represented by $A$. This is read, "$x$ is an element of $A$." $y \notin A$ means "$y$ is not an element of $A$."

EXAMPLE: In the case of the set $V = \{a, e, i, o, u\}$ mentioned earlier, $e \in V$ and $u \in V$ but $c \notin V$.

## Axiom S1

For every object $x$ and every set $A$, either $x \in A$ or $x \notin A$ exclusively.

The word *exclusively* is used to indicate that $x \in A$ and $x \notin A$ cannot both be true. We will use this expression consistently with the word *or*. When *or* is not followed by the word *exclusively*, it will mean that *both* statements connected by *or* may be true.

## Axiom S2

For every set $A$, each element of $A$ is distinct from the other elements of $A$.

## RELATED SETS

Suppose $F = \{a, b, c, d, e\}$ and $V = \{a, e, i, o, u\}$ as before. If we have made distinct symbols represent distinct objects, we find in comparing these sets that $F$ and $V$ have the elements $a$ and $e$ in common. We might thus have reason to consider a new set $\{a, e\}$. It might also be convenient to use the set composed of every different element in $F$ and $V$; that is, the set $\{a, b, c, d, e, i, o, u\}$, in which we are careful not to list the $a$ and $e$ twice. These two methods of selecting certain elements of two sets are so useful that it is appropriate to define them and to introduce some new symbols.

*Definition 1.3.* If $A$ and $B$ represent sets, then $A \cup B$ (read $A$ *union* $B$) represents the set obtained by combining or uniting sets $A$ and $B$. More precisely:
$$A \cup B = \{* \mid * \in A \text{ or } * \in B\}.$$

EXAMPLE: If $C = \{a, c, t\}$, $S = \{a, c, k, s\}$ and $T = \{o, p, t\}$ then
$$C \cup S = \{a, c, t, k, s\}$$
$$C \cup T = \{a, c, t, o, p\}$$
$$S \cup T = \{a, c, k, s, o, p, t\}$$

Observe that the letters $a$ and $c$ occur in both $C$ and $S$, but conforming to Axiom *S2* when listing the elements of the two sets in $C \cup S$, each letter is written only once. The special attention given to the elements $\{a, c\}$ in determining the elements of $C \cup S$ points out the utility of the concept defined next.

*Definition 1.4.* If $A$ and $B$ represent sets, then $A \cap B$ (read $A$ *intersec-*

*tion B*) represents the set of all elements common to $A$ and $B$. More precisely:

$$A \cap B = \{* \mid * \in A \text{ and } * \in B\}$$

It is not unusual for all the elements of one set to be elements of some other set. For example, every member of $C \cap S = \{a, c\}$ of the last example is also a member of $C$. In answering the last question of Exercise 1.1, you may have recognized the need to see that each name on one list was on the second list and that each name on the second list was on the first list.

*Definition 1.5.* If $A$ and $B$ represent sets, then $A \subseteq B$ (read $A$ is a *subset* of $B$) means that each element of $A$ is also an element of $B$. $A \subset B$ (read $A$ is a *proper* subset of $B$) means that $A \subseteq B$ but $A \neq B$.

Applying the notation of Definition 1.4 to the set $\{a, c\} = C \cap S$ of the preceding example reveals the relation $(C \cap S) \subset S$ since $a \in S$ and $c \in S$ while $(C \cap S) \neq S$ because $k \in S$ but $k \notin (C \cap S)$. When only two or three elements are said to belong to a set, as in the case of $c$ and $a$ being elements of $S$, we often write $a, c \in S$ instead of $\{a, c\} \subseteq S$.

It should also be apparent that when $A \subseteq B$ and $B \subseteq A$ are both true, then $A = B$. We have already noted an example of this situation by the two lists of names. $A \subseteq B$ would show that each name on list $A$ is also on list $B$, while $B \subseteq A$ says that each name on list $B$ is also on list $A$.

We have observed how certain elements of a set may be considered as a proper subset of that set. In examining sets $C = \{a, c, t\}$ and $S = \{a, c, k, s\}$ to determine the members of $C \cup S = \{a, c, t, k, s\}$ it should seem natural to list all the elements of $C$ or of $S$ first, as in

$$C \cup S = \{a, c, t, \cdots\}$$

Then those elements of $S$ which are not in $C$ must be added, to form

$$C \cup S = \{a, c, t, k, s\}$$

This corresponds to removal or subtraction of certain elements from a set. If the elements of $C = \{a, c, t\}$ were subtracted from $S = \{a, c, k, s\}$ the result would be the set $\{k, s\}$. Although there was no element $t$ to be removed from the set $S$, and all that was removed was the subset $\{a, c\}$, it is still customary to write $S - C = \{a, c, k, s\} - \{a, c, t\} = \{k, s\}$. The following statement formally defines this concept.

*Definition 1.6.* If $A$ and $B$ represent sets, then $A - B$ represents the set of all elements of $A$ which are not elements of $B$. More precisely:

$$A - B = \{* \mid * \in A \text{ and } * \notin B\}$$

EXAMPLE: If $G$ represents the set of all grapes in the world and $F$ the set of all fruit having seed, then $G - F$ would represent the set of all seedless grapes in the world, while $F - G$ represents the set of all fruit having seed except grapes.

*Exercise 1.2*

(In Problems 1–6, let $A = \{a, e, i, o, u\}$, $B = \{* \mid *$ is a letter of the word *bread*$\}$, and $C = \{n, e, s, w\}$.)

1. Use symbols to state whether the letter $a$ is or is not an element of $A$, $B$ and $C$.
2. Fill in the elements of $A \cap B = \{\qquad\qquad\}$
3. Fill in the elements of:

    (a) $A \cup B$ \qquad\qquad (b) $(A \cup B) \cap C$

4. Fill in the elements of:

    (a) $(A \cap B) \cap C$ \qquad\qquad (b) $A \cap (B \cap C)$

5. Fill in the elements of:

    (a) $A - B$ \qquad\qquad (b) $B - A$

6. Fill in the elements of:

    (a) $(A - B) \cup (B - A)$ \qquad\qquad (b) $(A - B) \cap (B - A)$

(In the following problems the letters $A$, $B$ and $C$ do not refer to the preceding sets.)

7. (a) Show that $A \subset B \rightarrow A \subseteq B$.
   (b) Does $A \subseteq B \rightarrow A \subset B$? Explain.
8. (a) If $x \in A$ and $A \subseteq B$, what else can be said about $x$?
   (b) $C \subseteq A$ and $A \subseteq B \rightarrow$ ?
9. Show that the following statements are not always true by giving a specific example in each case:

    (a) $A \subset (A \cup B)$ \qquad (b) $(A \cap B) \subset B$

10. Fill in the blanks to make valid implications:

    (a) $x \notin A \cup B \rightarrow x \underline{\qquad} A \underline{\qquad} x \underline{\qquad} B$.
    (b) $x \notin A \cap B \rightarrow x \underline{\qquad} A \underline{\qquad} x \underline{\qquad} B$.

## SPECIAL SETS

To illustrate two useful types of sets, consider the set $V$ of English vowels and the set $C$ of English consonants. Clearly $C \cup V$ is the English alphabet; but what is $C \cap V$? In an offhand manner, we might be inclined to say, "It is nothing." But we have agreed that the symbol $C \cap V$ shall represent a *set*—not *nothing*. Certainly, as a set, $C \cap V$ contains nothing since $C$ and $V$ have no elements in common. Therefore, for consistency and convenience, we speak of a *null* or empty set. Since all empty sets are equal (have the same elements) there is really only one null set.

At the other extreme, if all the sets used in a particular discussion were composed entirely of letters of the English alphabet, the set $C \cup V$ could serve as a *universal set*. Since the concept of the set of all elements (of all sets) leads to logical contradictions, we do not have a unique universal set. These ideas are combined in one definition:

*Definition 1.7.* The *null* set, $\emptyset$, is the unique set which contains no elements. A *universal* set, $U$, is the set of which every set in a given collection of sets is a subset. If $U$ is a universal set for a set $A$, then the set $U - A$ is called the *complement* of $A$ relative to $U$, sometimes represented by $A'$.

The fact that the intersection of the set $C$ of English consonants and $V$ of English vowels is the empty set gives rise to another language problem. Instead of saying that $C \cap V = \emptyset$, most people are inclined to say, "$C$ and $V$ do not intersect." While this is not necessarily a false statement, it does the same psychological damage as the statement "It is nothing;" that is, it tends to dismiss even the existence of $C \cap V$. The importance of this concept to mathematics should become quite clear, especially as we get into the development of number concepts. If such a relation between sets is important it might well be expected to have a name.

*Definition 1.8.* If $A$ and $B$ represent sets then $A$ and $B$ are *disjoint* iff (if and only if) $A \cap B = \emptyset$.

EXAMPLE: If a universal set $U$ is the set of all undergraduate students at your institution and $F$ represents the subset of all freshman students, then $F' = U - F$ represents the subset of all upper-classmen. Also, if $S$ represents the subset of all sophomores, then $F$ and $S$ are disjoint sets; i.e., $F \cap S = \emptyset$.

## VENN DIAGRAMS

A pictorial device which often proves helpful in learning about sets is the Venn diagram. As indicated, such a diagram is simply a picture in which each set under discussion is represented by an enclosed region. In Figure 1.1, the regions inside the two circles represent sets $A$ and $B$. The fact that the circles do not overlap shows that $A$ and $B$ are disjoint.

*Figure 1.1*

A Venn diagram is often drawn to show a universal set and the complement of a set. Usually, all the region inside a large rectangle represents the universe; then each set is a region inside the rectangle. In Figure 1.2, set $A$ is shown as the shaded region so that the unshaded region can be more clearly identified as $A'$, the complement of $A$.

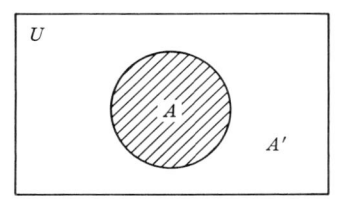

Sets

*Figure 1.2*

These diagrams are especially useful in demonstrating intersections, unions and subsets. Figure 1.3 with its key shows how the use of different shading or colors is effective for this purpose.

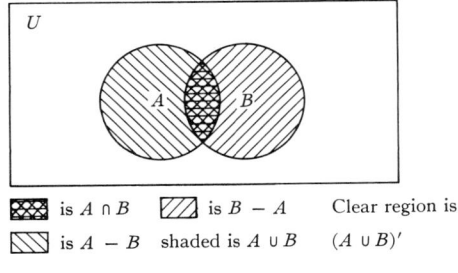

is $A \cap B$  is $B - A$   Clear region is
is $A - B$  shaded is $A \cup B$   $(A \cup B)'$

*Figure 1.3*

In general, the more complex the combinations of set symbols become, the more helpful Venn diagrams are in tracing the set represented, although this device is simply a visual aid and cannot be accepted as part of a mathematical proof. Notice, however, from Figure 1.3, how easy it is to pick out the region represented by the symbol $(A - B) \cup (B - A)$. Do you agree that it is the same region as $(A \cup B) - (A \cap B)$?

As a final note on Venn diagrams, you should know that some people draw the diagrams and then write symbols for the set elements in the appropriate regions. Figure 1.4 shows this method applied to $F = \{a, b, c, d, e\}$ and $V = \{a, e, i, o, u\}$.

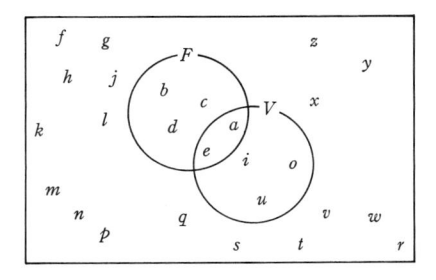

*Figure 1.4*

## THE USE OF SETS

Although certain other ideas concerning sets will be introduced in this text, we will wait to do so until they are needed to support other developments. In the next chapter, with the aid of what has been introduced thus far, we will develop the number concepts upon which our structure of arithmetic will be built.

*Exercise 1.3*

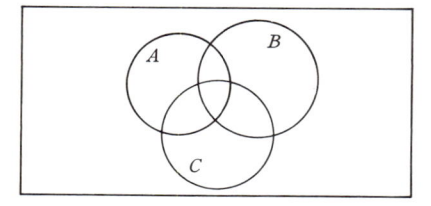

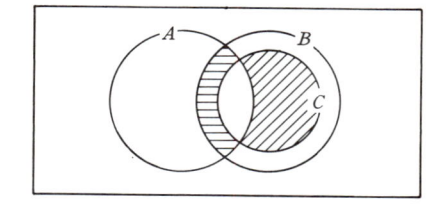

Figure 1

Figure 2

1. In Figure 1, shade in (or color) the region representing the set $A \cup C$.
2. Similarly, represent $(A \cap B) - C$.
3. In Figure 1, show the set $C \cap B'$.
4. Show in Figure 1 the set $(A \cup B)' \cap C$.
5. In Figure 2, what set is represented by the region shaded diagonally?
6. What set does the region shaded horizontally represent?
7. If $T$ is the set of all children under 10 years of age, $G$ the set of all girls under 7, and $B$ the set of all boys (all ages), draw a Venn diagram to represent these sets.
8. Draw Venn diagrams to compare $A - B$ and $A \cap B'$.
*9. Use Venn diagrams to decide whether $A \cup (B \cap C) = (A \cup B) \cap (A \cup C)$.
*10. Decide whether $A \cap (B \cup C) = (A \cap B) \cup (A \cap C)$.
11. Show that $A - B = A$ implies that $A$ and $B$ are disjoint sets.

# 2

# Cardinal and Ordinal Properties

## NUMBER

It is a curious paradox of the learning process that the most basic and elementary ideas are the most difficult to describe; and that we "learn" by imitating what other people say and do, but it is some time before the *significance* of what we have learned dawns upon us.

As an illustration of the first statement, just try to explain to someone what a number is. If you actually do try, you will probably appreciate more the significance of this chapter! As for the second statement, you probably learned to count (that is, call off the words "one, two,···") before you really understood what the word *two* meant.

When asked why they find mathematics difficult, most people say, "I guess it's because math is so abstract." Of course it's abstract! Any subject which deals primarily with ideas is abstract; and that includes just about every subject in school except those whose purpose is the development of physical skills. Some people believe, however, that much of the difficulty traditionally encountered in the study of mathematics is the result of a preoccupation with mathematical symbols to the exclusion of a proper understanding of the ideas which the symbols convey. In this text, careful distinction is made between the abstract ideas and the concrete symbols used to represent them, as an encouragement to the student to learn mathematics as a way of thinking in which special symbols are used to express thoughts, rather than as a mechanical shuffling of symbols on a page or on the chalkboard.

## ONE–ONE CORRESPONDENCE

The number concept can be grasped only through physical experience; and this is where sets play a role. The knowledge that two sets have the same number of elements is a more primitive concept than the knowledge of the number itself. Two children who cannot count can share equally a set of cookies by having one child take a cookie each time the other does. This pairing or matching of the elements of two sets is called a *one–one correspondence*. This is an unfortunate choice of words, since the word *one* is a symbol for a number. But this matching is done with no thought of the number one. Perhaps it might be more appropriately called an *element-to-element correspondence*.

*Definition 2.1.* There is a *one-one correspondence* between sets $A$ and $B$ (or between their elements) if and only if there is a set of pairs $(a, b)$ with $a \in A$ and $b \in B$ such that each element of $A$ is paired with an element of $B$, each element of $B$ is paired with an element of $A$, and no element of either set occurs more than once. $A \sim B$ (read *A is equivalent to B*), if and only if there is such a correspondence.

EXAMPLE: The sets $F = \{a, b, c, d, e\}$ and $V = \{a, e, i, o, u\}$ are equivalent, as shown by the pairings $\{(a, a), (b, e), (c, i), (d, o), (e, u)\}$ illustrated in Figure 2.1.

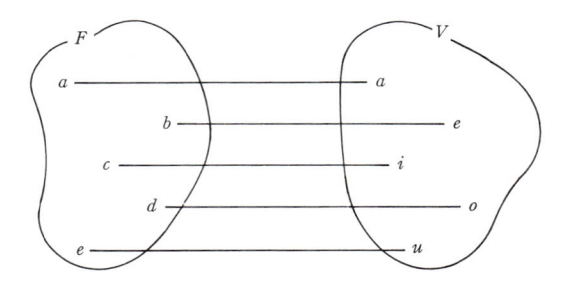

*Figure 2.1.* $F \sim V$; *A One–one Correspondence.*

Of course there are other pairings of the elements of $F$ and $V$, and the problem of devising a way to calculate the number of such pairings is a typical and interesting example of mathematical thought. However, we must examine the concept of number before we attempt to use the concept, and one-one correspondence or set equivalence is the means by which the number concept is comprehended.

*Definition 2.2.* The *cardinal number* of a set $A$ is that unique property which all sets equivalent to $A$ have in common.

In a practical way, this definition is simply saying that the only way anyone can understand the meaning of "two," for example, is to experience enough sets equivalent to $\{a, b\}$ that those sets could have no other common property.

When one perceives this property, he has abstracted the idea of *two-ness*. Most concepts are formed in this way during early childhood. Occasionally you will discover that you have a faulty or erroneous concept of something, which may mean that you stopped looking at examples too soon, when you thought you understood the idea!

## CARDINAL NUMBER AND NUMERALS

In the study of numbers our communication of ideas is hindered once again if we have no symbols to represent the ideas. So, we introduce the familiar Hindu-Arabic numerals for restricted usage, although they will be more thoroughly analyzed in the next chapter. Our purpose here is to invent these symbols, one by one, to represent cardinal numbers of sets, and then to use them as elements to build new sets.

We start with nothing. Then how can we have a set? Remember, there is a set requiring nothing—the null set. Let it be shown as { }, and let its cardinal number be called *zero* symbolized by the numeral 0. Now we have something—the symbol 0. Let's make a set with it: {0}. Since this set is not equivalent to the first set, { }, it must have a different cardinal number. Let's call it *one* and invent the numeral 1 to represent it. Thus we can build the set {0, 1} since 0 and 1 are distinct objects. But this set is not equivalent to either of the other sets, so we must invent a word and a numeral to represent its cardinal number. We will use *two* and 2. This process can be continued as long as new symbols can be produced. The first few steps of this device are summarized in Table 7.

### Table 7.
### *Cardinal Number Illustrated By A Recursion Procedure.*

| Set | Numeral | Word Symbol |
|---|---|---|
| { } | 0 | zero |
| {0} | 1 | one |
| {0, 1} | 2 | two |
| {0, 1, 2} | 3 | three |
| {0, 1, 2, 3} | 4 | four |
| . | . | . |
| . | . | . |
| . | . | . |

The method by which the finite cardinal numbers were developed also illustrates another very important property of these numbers; they have a natural *order*. They can be presented in other orders, but the order we have used is that which every child learns, and that which has the greatest practical value.

Although these numbers are commonly called cardinal numbers, they are also called *whole* numbers. We will use the latter term and refer to the cardinal and ordinal *properties* of these numbers.

*Definition 2.3.* The set symbolized by $W = \{0, 1, 2, 3, \cdots\}$ is the set of *whole numbers*. The set $N = \{1, 2, 3, \cdots\}$ is the set of *natural numbers*.

The punctuation symbol $\cdots$, called *points of suspension*, indicates the omission of similar material. If the symbol is used in a list having first and last elements, it is placed somewhere between these elements, as in $\{a, b, c, \cdots, y, z\}$. The reader would be expected to recognize this set as the complete English alphabet. In the case of the set $N = \{1, 2, 3, \cdots\}$ the points occur at the end of the list to indicate that the list goes on endlessly. Later, in our development of number systems, we will need suspension points at the beginning of a list because the set will have no first element.

## FINITE AND INFINITE SETS

Let us take a brief look at a property which is closely associated with the sets $N$ and $W$ just defined, and with the concept of cardinal number in general. Relative to their cardinal numbers, sets have been classified as either *finite* or *infinite* exclusively.

One of the simplest examples of one–one correspondence occurs when a set is compared with itself. For example, to show that $V \sim V$ where $V = \{a, e, i, o, u\}$, the most natural pairings would be $\{(a, a), (e, e), (i, i), (o, o), (u, u)\}$ where each element is paired with itself. Using the order concept, a second natural combination is the pairing of each element with the one which follows, as in $\{(a, e), (e, i), (i, o), (o, u), (u, a)\}$. In the last pair, to find a partner for $u$ it was necessary to go back and pick up the $a$, which had been omitted in the listing of second elements.

Application of this second method to the set $N$, however, illustrates a very different situation. From an examination of the pairings

$$\{(1, 2), (2, 3), (3, 4), (4, 5), \cdots\}$$

it becomes apparent that it will never be necessary to use the element 1 as a second partner in a pair. In other words, there seems to exist a one-one correspondence between the sets

$$\{1, 2, 3, \cdots\} \text{ and } \{2, 3, 4, \cdots\}$$

or in symbols previously introduced,

$$N \sim (N - \{1\})$$

Notice that the set $N - \{1\} = \{2, 3, 4, \cdots\}$ is a proper subset of $N$, so that $N$ is equivalent to one of its proper subsets.

*Cardinal and Ordinal Properties*

This property makes it possible to distinguish between a set which is equivalent to a proper subset of itself and a set which is not. Sets which have this property are called infinite sets, such as the sets $W$ and $N$. A set such as $V = \{a, e, i, o, u\}$ which has no equivalent proper subset is called *finite*. This property is not easy to use, and its illustration in the example above is not at all complete. We will examine later a logically equivalent definition more appropriate to this development.

*Exercise 2.1*

(Let $A = \{a, b, c\}$, $B = \{4, 3, 1, 5\}$, $C = \{p, t, s\}$, $D = \{\#, \$, *\}$, $E = \{c, e, o, u\}$.)

1. Show all pairs of equivalent sets as, for example, $C \sim D$.
2. Show that $C \sim D$ by (a) a picture; (b) a set of ordered pairs.
3. See how many different ways you can show $C \sim D$. (Two ways are different if they do not have exactly the same pairings.)
4. Are any of the sets $A, B, C, D, E$ above equivalent to set $N$ of Definition 2.3? Explain.
5. Give the cardinal number of each of the sets $A, B, C, D, E$ as (a) a word; (b) a numeral.
6. List five common situations which require an ordered sequence for successful operation.
7. How many basic sets of symbols can you think of which are used for keeping records in usable order?
*8. If possible, show that the set $T = \{2, 4, 6, 8, \cdots\}$ is equivalent to the set $N$ of natural numbers.
*9. Show that $W \sim N$.
*10. Find a set equivalent to the set $\{5, 10, 15, \cdots\}$. Show the one–one correspondence.

# COUNTING

A very important use of the *ordinal* property of the natural numbers $\{1, 2, 3, \cdots\}$ occurs in counting. Almost everyone knows how to count, but few people have bothered to analyze the significant aspects of the procedure. The following description of the counting process demonstrates the relation between the ordinal property of the natural numbers and the finite cardinal numbers.

Figure 2.2 illustrates a one–one correspondence between the elements of the set $V = \{a, e, i, o, u\}$ and those of the special set $\{1, 2, 3, 4, 5\}$. The order of the elements of $V$ is immaterial, but both the elements of $N = \{1, 2, 3, \cdots\}$ and the order of those elements is important in finding the cardinal number

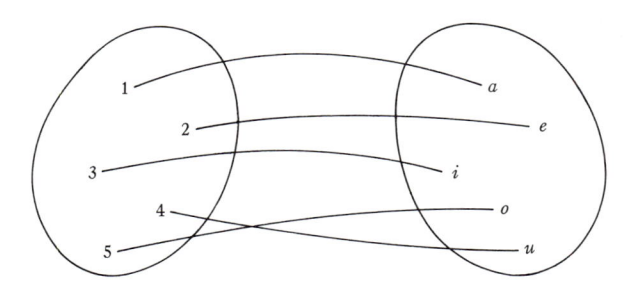

*Figure 2.2. Counting the Elements of V.*

of $V$. For example, the pairings of Figure 2.2 are given by the set

$$\{(1, a), (2, e), (3, i), (4, u), (5, o)\}$$

where the last natural number, 5, occurs in the last pair. For many purposes, the order

$$\{(2, e), (5, o), (3, i), (4, u), (1, a)\}$$

would be just as acceptable. However, it would be much more difficult to recognize the largest number of the set $\{1, 2, 3, 4, 5\}$ and to verify whether the correct set of numbers for counting the elements of $V$ had indeed been used. In other words, we do not want to imitate the mistakes of young children learning to count, who might say, "1, 2, 3, 6, 7, 4, $\cdots$."

In order to formalize the counting procedure and to clarify the following definition, we demonstrate in Table 8 the subsets of $N$, together with some new symbols for them. These sets are standard sets used to count the elements of other sets.

*Table 8. Standard Sets for Counting.*

| Set | Cardinal No. | Set Symbol |
|---|---|---|
| $\{\ \}$ | 0 | $N_0$ |
| $\{1\}$ | 1 | $N_1$ |
| $\{1, 2\}$ | 2 | $N_2$ |
| $\{1, 2, 3\}$ | 3 | $N_3$ |
| $\cdot$ | $\cdot$ | $\cdot$ |
| $\cdot$ | $\cdot$ | $\cdot$ |
| $\cdot$ | $\cdot$ | $\cdot$ |
| $\{1, 2, 3, 4, 5, 6\}$ | 6 | $N_6$ |
| $\cdot$ | $\cdot$ | $\cdot$ |
| $\cdot$ | $\cdot$ | $\cdot$ |
| $\cdot$ | $\cdot$ | $\cdot$ |

For example, $N_5 = \{1, 2, 3, 4, 5\}$ and since $V \sim N_5$, the cardinal number of set $V$ is 5. Similarly, $\{a, b, c\} \sim N_3$, so the cardinal number of set $\{a, b, c\}$ is 3.

*Definition 2.4.* Let $N_0 = \emptyset$ and $N_k = \{1, 2, \cdots, k\}$ for all other whole numbers $k$. Set $A$ is *finite* if and only if there is a set $N_n$ such that $A \sim N_n$; $n$ is the *cardinal number* of $A$. A set which is not finite is said to be *infinite*.

EXAMPLE: From Definition 2.4, $N_0 = \{\ \}$ and 0 is its cardinal number. The cardinal number of the set $\{*, \#\}$ is 2, since, by Definition 2.4, $N_2 = \{1, 2\}$ and $\{*, \#\} \sim N_2$ by the pairings $\{(*, 1), (\#, 2)\}$. The emphasis of Definition 2.4 is on finite counting procedure, which is the basis of elementary arithmetic.

## RELATIONS OF SETS AND CARDINAL NUMBERS

Thus far, we have seen two ways in which sets can be related—by set *equality* and by set *equivalence*. You should realize that when two particular sets are equal (have the same elements) they are also equivalent (there is a one–one correspondence of elements). But, two sets which are equivalent are not necessarily equal. However, the equivalence of two sets implies that they have the same cardinal number, according to Definition 2.2. We shall employ our practical definition, 2.4, in the following example to show that these two definitions agree.

EXAMPLE: For the sets $F = \{a, b, c, d, e\}$ and $V = \{a, e, i, o, u\}$ we saw that $F \sim V$ by the pairings

$$\{(a, a), (b, e), (c, i), (d, o), (e, u)\}$$

In our example of counting, we also saw that $V \sim N_5$ by the set of pairs

$$\{(a, 1), (e, 2), (i, 3), (o, 4), (u, 5)\}$$

It should now be easy to see that a one–one correspondence between $F$ and $N_5$ is established by replacing each element of $V$ in the second set of pairs by the element of $F$ which is paired with it in the first set, from which

$$\{(a, 1), (b, 2), (c, 3), (d, 4), (e, 5)\}$$

so that $F \sim N_5$. This method of proof can be used to demonstrate the *transitive* property of set equivalence; that is, if $A \sim B$ and $B \sim C$ then $A \sim C$ also.

## NON-EQUIVALENT SETS

Since two sets selected at random are not likely to be equivalent, we need some means of describing more accurately how such sets are related. Suppose,

for example, that we wish to compare the cardinality of sets
$$A = \{a, b, c, d, e\}$$
and
$$B = \{x, y, z\}$$
(These sets have been purposely selected so that $A \cap B = \emptyset$, in order to avoid confusion.) To answer the question, "Is $A \sim B$?" we would examine pairings, such as
$$\{(a, x), (b, y), (c, z), (d, \quad), (e, \quad)\}$$
and note that $B$ does not have enough elements to complete the pairings. Since a list of incomplete pairings is awkward, we seek another way to describe the situation.

If we examine the completed pairings
$$\{(a, x), (b, y), (c, z)\}$$
we note that $A$ has a proper subset
$$C = \{a, b, c\}$$
for which $B \sim C$. Thus we might observe that if $A$ is not equivalent to $B$ (written $A \not\sim B$) and there is $C \subset A$ for which $C \sim B$, then "the cardinal number of $A$ is greater than the cardinal number of $B$." The following definitions will formalize these observations.

*Definition 2.5.* If $A$ represents a set, then $n(A)$ represents its cardinal number. $n(A) > n(B)$, read "The cardinal number of $A$ is greater than the cardinal number of $B$," iff $A \not\sim B$ and there is $C \subset A$ for which $C \sim B$.

At first glance it might seem impossible for sets $A$, $B$ and $C$ to exist such that $B \sim C$, $C \subset A$ and $B \sim A$ also. In other words, it might seem unnecessary in Definition 2.5 to stipulate that $A \not\sim B$, in addition to demanding that $C \subset A$ and $C \sim B$. However, where $E = \{2, 4, 6, 8, \cdots\}$, $E \sim N$. Also, for $Q = \{1, 3, 5, 7, \cdots\}$, $Q \subset N$ and $Q \sim E$. (Let each odd number $n$ correspond to the even number $n + 1$.) It turns out that this peculiar set of relations can happen only with infinite sets. In fact, as we have indicated, some people define a set to be infinite when it is equivalent to one of its proper subsets, as, for example, $N$ is infinite since $N \sim E$ and $E \subset N$. The following exercises should help clarify these ideas.

## Exercise 2.2

1. Draw a picture similar to Figure 2.2 to show how to count the elements of the set $\{b, d, h, k, l, t\}$.
2. Pair each element of the set $H = \{b, d, h, k, l, t\}$ with the first elements of $N$. From these pairings, identify the standard set $N_k \sim H$.

## Cardinal and Ordinal Properties

3. Use the form $N_k$ to indicate the respective standard set equivalent to:

   (a) $\{a, b, c, d\}$    (b) $\{x\}$    (c) $\{\emptyset\}$

4. Exhibit sets equivalent to:

   (a) $N_2$    (b) $N_7$    (c) $N_9 - N_4$

5. If $A = \{a, b, c, d\}$, use symbols to show the relation between $n(A)$ and the cardinal numbers for:

   (a) $B = \{1, 2\}$    (c) $D = \{1, a, 2, b, c\}$
   (b) $C = \{\ \}$    (d) $F = \{f, o, u, r\}$

6. Find a proper subset of $S = \{v, w, x, y, z\}$ and show by Definition 2.5 that $n(S) > n(T)$ where $T = \{p, q\}$.

7. Use Definition 2.5 and sets of your choice to prove that $6 > 4$.

(In Problems 8–10 use your knowledge of addition, although we have not yet stated a definition.)

*8. If $A = \{a, b, c\}$ and $B = \{d, e\}$, how do $n(A \cup B)$ and $n(A) + n(B)$ compare?

*9. If $A = \{a, b, c\}$ and $B = \{c, d\}$, how do $n(A \cup B)$ and $n(A) + n(B)$ compare?

*10. For the sets of Problem 9, how do $n(A) + n(B)$ and $N(A \cup B) + n(A \cap B)$ compare?

11. Find a proper subset of $E = \{2, 4, 6, 8, \cdots\}$ which is equivalent to $E$.

## ADDITION OF WHOLE NUMBERS

In its most elementary phase, addition is the application of counting to the problem of finding the cardinal number of the union of two sets. We demonstrate this application when we say to a child, "How much is 3 fingers and 2 fingers?" At a later age he is asked "How much is $3 + 2$?" as we seek to abstract the number idea from the sets of concrete objects. The combining of number and set, as in the former statement, has caused some people to talk about *concrete numbers*, encouraging a common misconception of number. Set language enables us to point out the distinction between the *objects* and the *number* of objects. We no longer add 2 apples and 3 apples; instead, we form the union of two sets of apples and find the cardinal number of that set, initially, by counting.

For example, suppose $A = \{a, b, c, d\}$ and $B = \{x, y, z\}$. Then $A \cup B = \{a, b, c, d, x, y, z\}$, so that $n(A \cup B) = 7$. Since $A$ and $B$ are disjoint sets, each element of set $A$ and each element of set $B$ contributes to the total number of elements of set $A \cup B$, so we would like to say that

$$n(A) + n(B) = n(A \cup B)$$

or

$$4 + 3 = 7$$

If the two sets had not been disjoint, their union could not have been used so conveniently. For example, if $A = \{a, b, c, d\}$ and $C = \{b, d, h\}$ then $n(A) + n(C) = 4 + 3$, just as in the sum $n(A) + n(B)$ above. However, since $A \cup C = \{a, b, c, d, h\}$, $n(A \cup C) = 5 \neq n(A) + n(C)$. To use the union of two sets to define the sum of their cardinal numbers, it is important to specify that the sets be disjoint.

*Definition 2.6.* For disjoint sets $A$ and $B$, $n(A) + n(B) = n(A \cup B)$ and is called the *sum* of $n(A)$ and $n(B)$.

Such a definition provides a way of identifying the sum of two cardinal numbers. However, a definition alone does not guarantee that the object which it defines actually exists. In other words, we would like to be assured that the sum of any two whole (cardinal) numbers exists. Of course, we are already convinced that it does, but we would like to have a way to show it.

For example, suppose $m, n \in W$. How can we show that $m + n$ is actually meaningful? According to Definition 2.6, if we can produce two disjoint sets of cardinality $m$ and $n$ respectively, then the cardinal number of their union defines $m + n$. Now Definition 2.4 identifies $N_m$ and $N_n$ as sets of cardinality $m$ and $n$. However, these sets are not disjoint unless either $m = 0$ or $n = 0$; thus, we need to create sets equivalent to $N_m$ and $N_n$ which *are* disjoint. An idea which we have used in our discussion of one–one correspondence enables us to do this in a very natural way. Let $A = \{a\}$ and $B = \{b\}$, so that $A$ and $B$ are disjoint. Then

$$C = \{(a, 1), (a, 2), \cdots, (a, m)\}$$

and

$$D = \{(b, 1), (b, 2), \cdots, (b, n)\}$$

are certainly disjoint, since no element (ordered pair) of one set is found in the other. Thus $n(C \cup D) = m + n$, so we have outlined a proof for the following important theorem.

## Theorem 2.1

For all $m, n \in W$ the sum $m + n \in W$.

# SUBTRACTION OF WHOLE NUMBERS

Although we shall define subtraction later, in a manner more appropriate to generalization, we would violate our promise to make mathematics meaningful if we ignored the intuitive concept of *take away* which relates set subtraction and subtraction of whole numbers.

The concept of subtraction of sets is most natural when one set is a subset

of the other. Thus if
$$A = \{a, b, c, d, e\}$$
and
$$B = \{a, b\}$$
then $A - B = \{c, d, e\}$ as pictured in Figure 2.3. This is the concept of sub-

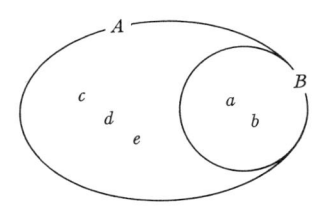

**Figure 2.3.** $A - B$ where $B \subset A$.

traction which is usually introduced to children, with the observation that
$$n(A) - n(B) = n(A - B)$$
or
$$5 - 2 = 3$$

It should be observed that the sets $A$, $B$ and $A - B$ are also related by the statement
$$(A - B) \cup B = A$$
Since $(A - B) \cap B = \emptyset$, it also follows that
$$n(A - B) + n(B) = n(A)$$
or
$$3 + 2 = 5$$

This relation between the operations of addition and subtraction has been used to define subtraction by means of addition, and to check the results of subtraction. For the present, however, we shall limit ourselves to the concept of subtraction of whole numbers by means of set subtraction.

The restriction of subtraction to the use of sets limits subtraction to the case
$$m - k = n(A) - n(B)$$
where either $m > k$ or $m = k$, since $B \subseteq A$ is required. However, it is not necessary to demand that $B \subseteq A$ in order to define $n(A) - n(B)$, for $n(A) = n(B)$ implies $A \sim B$, and $n(A) > n(B)$ implies the existence of $C \subset A$ for which $C \sim B$. Thus if $n(A) \geq n(B)$, there is a set $C \subseteq A$ such that $n(C) = n(B)$, so that
$$n(A) - n(B) = n(A) - n(C)$$

and the subtraction can be calculated as before by the cardinal number
$$n(A - C)$$
For example, if $A = \{a, b, c, d, e\}$ and $B = \{x, y, z\}$, let $C = \{a, b, c\}$. Then $B \sim C$ and $A - C = \{d, e\}$. Thus
$$n(A) - n(B) = n(A) - n(C) = n(A - C)$$
or
$$5 - 3 = 5 - 3 = 2$$

## Exercise 2.3

1. Use $B = \{b, c, d\}$ and $V = \{a, e, i, o, u\}$ to add $3 + 5$, by Definition 2.6.
2. Using Definition 2.6, form appropriate sets to show that $3 + 0 = 3$.
*3. Using Definition 2.6, prove that $k + 0 = k$ for every whole number $k$.
*4. Use $A = \{a, b, c, d\}$, $B = \{b, d, h\}$ and other sets of your choice to demonstrate that for general sets, $A, B, n(A) + n(B) = n(A \cup B) + n(A \cap B)$.
5. Draw a picture of sets to illustrate the sum $2 + 3$.
6. Could the relation $n(A) + n(B) = n(A \cup B) + n(A \cap B)$ be used as a definition for $n(A) + n(B)$ instead of Definition 2.6? Explain.
7. Select a set to use with $A = \{a, b, c, d, e, h\}$ to demonstrate $6 - 3$. Make the demonstration.
*8. Prove that $k - k = 0$ for every $k \in W$.
9. Is it generally true for any sets $A$ and $B$ for which $n(A) > n(B)$ that that $n(A - B) + n(B) = n(A)$? Illustrate.
10. Try to prove that $n(A \cup B) - n(B) = n(A - B)$, or show that it is not true.

# 3

# Numerals and Bases

No one actually knows the nature of the first device used by men to represent numbers. During the past few centuries, however, in the process of world exploration, a number of primitive cultures have been discovered which still exist in this modern age. From these people and from records of ancient civilizations found in archeological diggings, we can gain some idea of the evolution of numeration.

## PREHISTORIC PERIOD

It seems to be a generally accepted maxim that man's inventions have evolved as the natural result of recognized needs. Consequently, it seems unlikely that man would have invented symbols for numbers until his business transactions demanded some precision and involved so many objects that he could no longer identify them by sight. Thus, for example, when a shepherd had so many sheep that it was no longer feasible to recognize them individually to see that all were safe in the fold, he might have selected a bag of pebbles or a bundle of sticks equivalent to the set of sheep. If the sheep and pebbles are still equivalent at day's end, then the sheep are all in the fold. (Some professors are known to check class attendance by this method!)

When writing was invented, the shepherd probably counted his sheep with tally marks, such as

//////////////

To make it easier to distinguish one tally mark from another, they were eventually grouped, perhaps like this:

///// ///// ////

or even like this:

⌁⌁⌁ ⌁⌁⌁ ////

which may have evolved eventually to the symbols:

V  V  IIII

Such symbols (not words) used to represent numbers are called *numerals*.

## EGYPTIAN NUMERATION

With the discovery of a document called "The Rhind Papyrus," we have acquired fairly good knowledge of the extent of mathematical development in ancient Egypt. Our concern for the present, however, is the manner in which they represented numbers.

As we might expect, the Egyptian method of numeration was relatively primitive. They did improve on the tally system, however, by introducing the method of replacing ten tally marks with a single symbol. A natural generalization, then, was to replace ten of the new symbols by a still different symbol, and so on. Table 9 shows the numbers with their symbolic representations.

*Table 9. Egyptian Numerals.*

| Number | Symbol |
|---|---|
| one | │ |
| ten | ∩ |
| hundred | ℗ |
| thousand | ⚘ |
| ten-thousand | ⟨ |
| hundred thousand | ⌒ |
| million | 𓀀 |

## Numerals and Bases

EXAMPLE: Expressing a number in Egyptian hieroglyphics is somewhat like giving a customer change with a purchase. Just as the customer is given a minimum number of pieces of money, so the Egyptian numeral employs the least number of symbols possible. Even so, writing a *numeral* for nine hundred ninety-nine becomes quite a laborious process—try it! I shall write something short, like three hundred twenty-seven.

*Exercise 3.1*

1. What numbers are represented by the following symbols?
   (a) [hieroglyphic symbols]
   (b) [hieroglyphic symbols]

2. Express in Egyptian numerals:
   (a) 77
   (b) 2124
   (c) 40,404

## GREEK AND ROMAN NUMERATION

The Roman contribution to the concept of numeration amounted to little more than a refinement of the basic work of the Egyptians. The number of repeated symbols was reduced by two devices: (1) they introduced intermediate symbols, such as V for five so that no more than four tally marks were ever needed, as in IIII for four or VIIII for nine; and (2) they had a standard order for different symbols (highest valued symbols to the left) and by reversing the order of a pair of symbols, they could show subtraction. For example, VI uses the symbols in standard order, meaning $5 + 1$, or 6, while IV has the higher valued symbol on the right, and means $5 - 1$, or 4. Thus, four can be written with only two symbols instead of four tally marks.

Roman numerals form a crude system and computation with them is quite laborious. The following table, rules and examples are given only because the influence of Roman civilization on the Western World has caused us to retain the use of Roman numerals in several places. Perhaps someday our society will banish both Roman numerals and the awkward English system of measure.

### Table 10. Roman Numerals.

| Number | Roman Numeral |
|---|---|
| one | I |
| five | V |
| ten | X |
| fifty | L |
| hundred | C |
| five hundred | D |
| thousand | M |

EXAMPLE: Since two of any intermediate symbols (V, L or D) are equivalent to one of the basic symbols (X, C or M), the intermediate symbols are never repeated; neither are they subtracted. Thus X, C and M are used rather than VV, LL and DD. Similarly, there is no need to write a numeral such as LC (100 − 50) since this is more easily represented by L. On the other hand, it is much easier to write CD for four hundred than CCCC.

## Exercise 3.2

**1.** Express the following in Roman Numerals:

  (a) 342
  (b) 449
  (c) 1551

**2.** What numbers are represented by:

  (a) LXIV
  (b) XLVI
  (c) MCMLXV

Although the Greek system of numeration has had no apparent effect on modern mathematical notation, it initiated the second major improvement in numeration. The Greeks went to the extreme of having a different single symbol for each of the first twenty-four numbers, by the simple expedient of using the twenty-four letters of their alphabet. Of course there are some obvious drawbacks to such a system, so we shall simply give you a table of part of their system and then proceed to the system which we use today.

*Numerals and Bases*

### Table 11. Greek Numerals.

| Number | Numeral | Name |
|--------|---------|------|
| one | $\alpha$ | alpha |
| two | $\beta$ | beta |
| three | $\gamma$ | gamma |
| four | $\delta$ | delta |
| five | $\epsilon$ | epsilon |

## HINDU-ARABIC NUMERALS

Although there are a number of interesting variations on the numeration principles which we have studied and are about to discuss, we wish to focus attention on these ideas in the context of our own system of numeration. If you are interested in further reading along this line, consult the bibliography or your instructor for references.

As its name implies, there is some doubt as to the true origin of this system of numeration, since at the time it was introduced in Europe, it was used by both the Hindus and the Arabs. The symbols used then were quite different from ours today, and varied greatly from region to region, and person to person, since there was no printing press to help standardize them. It is interesting to note that the symbols 1, 2, 3, 4, 5, 6, 7, 8, and 9 were in use long before the symbol 0 for zero was adopted.

For some reason, perhaps the fact that man has ten fingers, ten basic symbols (digits) were invented to represent the first ten elements of $W$. Because it is based upon ten, this system is called the *decimal system* of numeration, from the latin prefix *deci*, meaning ten. In this respect the decimal system is like the Greek system, but it does not have the disadvantage of using a great many symbols which are also used for another purpose.

## PLACE VALUE

The remarkably ingenious feature of the Hindu–Arabic system of numeration which makes it so far superior to others we have mentioned, is the device which allows representation of extremely large numbers with the use of only ten different symbols. Indeed, as we shall soon demonstrate, the same numbers, by the same principle, could be represented by using as few as *two* different symbols. The germ of this concept was recognized by the Romans when they arranged the individual symbols in a particular relative order. However, they did not make full use of this idea, probably because they also had different symbols to represent very large numbers, whereas the largest number which can be represented by a single symbol in the Hindu–Arabic system is nine. This concept is that of *place value*, and means simply that the number represented by a *digit* (one of the single symbols) depends upon its *position* relative to the other digits in the numeral.

EXAMPLE: In the decimal numeral 2035 the digit 5 represents the number five, the digit 3 represents the number thirty, the digit 0 represents the number zero (always), and the digit 2 represents the number two thousand. The complete numeral then represents the sum of these numbers, two thousand thirty-five.

The Babylonians used the idea of place value to a limited extent hundreds of years earlier, but they lacked the second clever idea which makes the place value system so functional. In the example above, since the zero contributes nothing to the sum it might appear to be useless. Without the zero, however, the numeral would appear as 235, in which case it could not be distinguished from 2305 or 2350 or even 20035. The Babylonians had this problem with numeration. Strangely enough, the idea of a symbol for zero was the most sophisticated, and hence the most recent, member of the evolution of basic concepts which finally produced an efficient numeration system. Let us pause and collect these ideas and terms into formal definitions.

*Definition 3.1.* A *numeral* is a written non-language symbol which represents a number. A *numeration system* is a procedure which establishes numerals for a set of numbers.

*Definition 3.2.* A *place value* numeration system has the following properties:

(1) There is a set of at least two basic numerals called *digits*.
(2) The digits represent the first elements of $W$ in their natural order.
(3) Each number can be represented by a standard numeral which consists of a (usually *horizontal*) row of digits.
(4) The relative position (*place*) of each digit in a standard numeral helps to determine the number which it represents. The number represented by the digit *one* in any place is called the *place value* of that place.

EXAMPLE: In order to understand better the rather involved definitions above, let us examine the number two thousand thirty-five of our previous example. Since 2035 is a symbol for that number and is not a word or language symbol, it is a numeral by Definition 3.1. This numeral is a row of digits, since 2, 0, 3, and 5 are elements of 0, 1, 2, 3, 4, 5, 6, 7, 8, 9, which represent the first ten elements of $W$ in natural order. Also, (since there are ten), there are at least two digits. Finally, 2 is in thousands place, since 1 in that position would represent a thousand; 0 is in hundreds place, since 100 represents a hundred; the place value of the place occupied by 3 is ten since 10 represents ten; and 5 is in one's place, since 1 represents one. Thus, 2035 represents the number 2 thousands, 0 hundreds, 3 tens and 5 ones, or two thousand thirty-five.

*Exercise 3.3*

1. We use sets of numbers daily in various place value systems. For example, 9/16/65 is a calendar date. The numbers 36-26-36 might represent a beauty contestant's measurements. Give two other examples of place value systems.
*2. Write the place value of each digit in:
   (a) 27    (b) 385    (c) 40619
*3. Write the number represented by each digit in Problem 2.
4. Some place value systems employ a different set of digits in different places. Thus if 2/1/7 measures yards, feet and inches, the 7 is one of twelve digits, 1 is one of three, and 2 is one of five ($5\frac{1}{2}$ yards is a rod). Look up the British pound-shilling-pence monetary system.

## PLACE VALUE SYSTEMS

We have both stated and implied in the definitions that the only necessary restriction on a set of digits is that it contain symbols for at least the numbers zero and one, while any greater number of digits is permissible. We can demonstrate this principle and analyze the general nature of place value numeration systems by studying non-decimal place value systems.

*Definitions 3.3.* The number of digits in a place value system is called the *base* of the system. A place value system having $k$ digits is called simply a *base k* system.

The decimal system is also called a base ten system. Conversely, systems of base two, base three, and base twelve are called respectively *binary*, *ternary* and *duodecimal* systems. Certain other bases are given special names, but we shall be concerned only with the mathematical properties common to all such systems. Our purpose is to derive a thorough understanding of our decimal system and the way it works.

## NON-DECIMAL BASES

Every whole number can be represented in a place value system regardless of the base (provided it is at least two) or the types of symbols used (as long as they are distinguishable). We will illustrate this fact by an example using bases two and seven, and by another example using base twelve. To emphasize the difference between the number and the numerals used to represent it, we will exhibit a set of objects to be counted and then show how to find the numerals, representing its cardinal number in both base seven and base two. To show the base used for a particular numeral, we write the word name for the base as a subscript. For example,

$$2031_{five}$$

is read "Two zero three one, base five." Notice that even though we use the familiar Hindu-Arabic symbols as digits and call them by their usual names, we do not read the numeral in this example as two thousand thirty-one, because the 2 does not represent two thousand when the base is not ten. This idea will become clearer as we demonstrate the place values in our examples. Although we are free to invent other symbols to use as digits, it will be easier to use the standard digits for the first elements of $W$, and invent additional symbols as needed for bases greater than ten.

EXAMPLE: There are fifteen asterisks in the array below—too many to be represented by any one of the seven digits 0, 1, 2, 3, 4, 5, 6 of a *base seven* system. Therefore, we group them in subsets of seven, as shown:

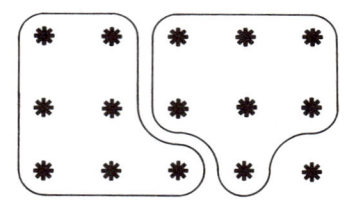

The remaining asterisks are less than seven, so their number can be represented by a digit—in this case, 1. The number of groups is also less than seven, so it can be represented by the digit 2. Then, by writing these digits in a row with place values increasing toward the left, we obtain the numeral $21_{seven}$. Notice that the position to the right has place value one, since its digit counts the number of single, or ungrouped, elements; the next place has a place value which is the base (seven in this case) since each of the groups which its digit counts is seven. Thus $21_{seven}$ is interpreted as

2 groups of seven plus 1 one

or

2 sevens + 1 one

which we sometimes write in our more comfortable decimal system as

$$2 \times 7 + 1 \times 1$$

To maximize understanding of this important principle, and to relate this treatment to most of the available literature on expanded notation, we introduce multiplication as an operation already familiar to the reader. We are thus deviating from the logical order of development, and the student should recognize that this form would not be used to explain the significance of place value in the addition algorithm to young children who have not studied multiplication.

The above methods for expressing $21_{seven}$ are examples of *expanded* notation, meaning that each digit is shown separately and multiplied by its place value so that the sum is then the number represented. Thus $2 \times 7 + 1 \times 1$, for example, is fifteen. If the same method is used and all numerals written in base seven, an interesting result is obtained. You see, the number seven written as a base seven numeral is $10_{seven}$ that is, 1 seven and 0 ones. Therefore, the numeral $21_{seven}$ written in expanded notation in base seven symbols is

$$2 \times 10_{seven} + 1 \times 1_{seven}$$

or, dropping the word *seven*,

$$2 \times 10 + 1 \times 1$$

If you were to compute this number, even though you thought it to be in base ten, you would get the numeral 21.

## Exercise 3.4

1. Write word expressions for the following numerals as you would read them aloud:
   (a) $3145_{six}$  (b) $2003_{four}$

2. Write numerals representing the number of asterisks shown in the set in the following bases: (a) eight, (b) five, (c) six.

3. Suppose the symbols $*$, $\$$, $:$, $)$, $\#$ constitute the complete set of digits for a place value system, representing the numbers zero through four successively.
   (a) What is the numeral for thirteen?
   (b) What number would $)\#$ represent?

4. Write the following numerals in expanded notation using *words* for place value.
   (a) $52_{six}$  (b) $47_{nine}$

5. Write the following numerals in expanded notation using decimal notation for place value. Find the number represented in each case (i.e., write a decimal numeral for it).
   (a) $37_{eight}$  (b) $50_{seven}$

EXAMPLE: Now let us use another set of fifteen asterisks to try to find a numeral in base two, using only the two digits 0, 1. Following the procedure of the previous example, we group them by two's, as illustrated:

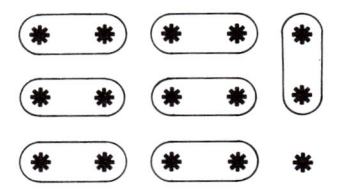

Again we notice that the number of the asterisks remaining is less than the base, and hence can be represented by a digit, **1**. The number of groups, however, is not less than the base, so cannot be expressed by a digit. We seem to be faced with a new problem, but it is actually the same problem we had when there were too many asterisks to be expressed by a digit—we just have groups instead of asterisks. So, we should be able to solve this problem by the same method. We shall simply combine the groups in pairs, as indicated below.

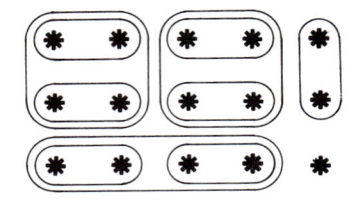

These larger groups might be called bi-groups, since they result from two groupings. Now we can represent both the left-over asterisk and the left-over group by digits—each by the digit 1. However, we still have too many bi-groups to represent by a digit, so it will be necessary to make some tri-groups, as illustrated here:

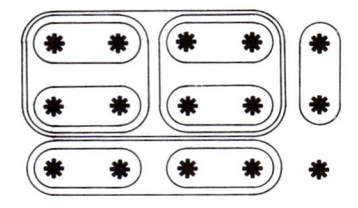

Are we finished? We have 1 asterisk, 1 group, 1 bi-group, and 1 tri-group; each set can now be represented by a digit. Since our checklist of sets was carefully arranged in order of size, to write the numeral we need only start at the right with the first digit and proceed to the left with each successive digit. Thus, $1111_{two}$ represents the number fifteen.

Notice how the place values are related. From right to left they are one, two, four, and eight. Each is twice the place value of the place to its right. This should not surprise us, since each new group of asterisks was composed of two of the preceding groups. Thus the group had 2 asterisks, the bi-group had 2 × 2, and the tri-group had 2 × 2 × 2.

Examining the number represented by $1111_{two}$ in expanded notation, we have

1 eight + 1 four + 1 two + 1 one

or

1 × 8 + 1 × 4 + 1 × 2 + 1 × 1

or

1 × (2 × 2 × 2) + 1 × (2 × 2) + 1 × 2 + 1 × 1

or

$1 \times 2^3 + 1 \times 2^2 + 1 \times 2^1 + 1 \times 2^0$

where the notation $2^3$ represents the number found by multiplying three 2's together. It is a device (which you may remember from Algebra) for expressing repeated multiplication concisely.

## PLACE VALUE PRINCIPLE

Let the letter $b$ represent any base (number of digits). The preceding examples illustrate the following important principle concerning the place values in a standard numeral in base $b$.

First place has place value 1

Second place has place value $b$

Third place has place value $b^2$

Fourth place has place value $b^3$

...

$n^{th}$ place has place value $b^{n-1}$

For example, the numeral $1234_{five}$ represents, in decimal notation, the number

$1 \times 5^3 + 2 \times 5^2 + 3 \times 5^1 + 4 \times 5^0$

or

1 × (5 × 5 × 5) + 2 × (5 × 5) + 3 × 5 + 4 × 1

or

1 × 125 + 2 × 25 + 3 × 5 + 4 × 1

or

125 + 50 + 15 + 4 = 194

This illustrates how an understanding of the place value system of numeration enables us to convert a numeral in any base to a decimal numeral, so we can more easily recognize the number.

## Exercise 3.5

1. Write numerals representing the number of asterisks shown in the set in the following bases:
   (a) four
   (b) three
   (c) two

   ```
   * * * * *
   * * * * *
   * * * * *
   * * * * *
   ```

2. Let the symbols *, $, :, ), # represent the numbers zero through four, as a complete set of digits.
   (a) What number does : *# represent?
   (b) Write the numeral for eighty-seven.

3. Write the following numerals in expanded notation using words for place value.
   (a) $312_{four}$   (b) $2012_{three}$   (c) $10110_{two}$

4. Write the numerals of Problem 3 in expanded form with place values expressed in decimal (base ten) numerals.

5. Write the numerals of Problem 3 in expanded form with place values expressed as powers of the base (in decimal notation).

6. Write the numerals of Problem 3 in expanded form with place values expressed in the base indicated in the particular problem.

7. Write the numerals of Problem 3 in expanded form with place values expressed as powers of the base (in the numerals of that base).

## A SHORT METHOD

The procedure to find the numeral in a given base to represent a given number, as demonstrated in the preceding examples, is quite lengthy and depends upon having a set of concrete objects at hand. Surely we should be able to devise an abstract means of performing the operation, by using our familiar decimal system. Let us analyze the base two example.

Given the number fifteen, we can express it as 15 in the decimal system. We can also write the base two as 2. Although we have not progressed to that point in our development of arithmetic in this course, we will cheat a little and assume that we can divide 15 by 2 to find that we have seven groups of 2's, and 1 left over. The *algorithm* (or procedure) best suited to this purpose is one commonly used for short division, as shown below:

$$\begin{array}{r} 2\underline{|15\phantom{0}} \\ 7 + 1 \text{ element} \end{array}$$

Similarly, the 7 groups combine to form 3 bi-groups with 1 group left over:

$$\begin{array}{r} 2\underline{|7\phantom{0}} \\ 3 + 1 \text{ group} \end{array}$$

*Numerals and Bases* 49

and the 3 bi-groups become 1 tri-group with 1 bi-group left over, thus obtaining the numeral

$$1111_{two}$$

as before. The algorithm of short division makes it possible to perform these steps compactly, as shown below. If the place values are not written with the

```
2| 15                                          ones
  2| 7                           twos  + 1 one
    2| 3            fours + 1 two  + 1 one
       1 eight + 1 four + 1 two  + 1 one
```

remainders and the remainders are not repeated each time, the algorithm is extremely compact. Continuing the division until a zero quotient is obtained

```
2| 15      Remainder
  2| 7        1
    2| 3      1
      2|1     1
         0    1
```

places all the digits in a column as they should appear from right to left in the numeral.

EXAMPLE: For a final example of a numeral in a non-decimal base, we shall examine the base twelve. Since twelve digits are needed in base twelve, we will need all ten digits of our decimal system, and still other digits to represent the numbers ten and eleven. It is customary to use the letters *t* and *e* for this purpose, obtaining the set of digits {0, 1, 2, 3, 4, 5, 6, 7, 8, 9, *t*, *e*}.

Suppose we choose to express the number four hundred twenty-five in base twelve. Since short division by 12 is not easy, you may need pencil and paper to check the work below.

```
12| 425       digit
  12| 35        5
    12| 2       e (eleven)
        0       2
```

Thus, $425_{ten} = 2e5_{twelve}$, or in expanded notation,

$$2 \times (12)^2 + e \times 12 + 5 \times 1$$

or

$$2 \times 144 + 11 \times 12 + 15 = 288 + 132 + 5 = 425$$

in base ten, which checks our result.

## COUNTING

As indicated in Chapter 2, there are two meanings for the verb *count*. In that chapter, we were primarily concerned with counting as a means of determining the cardinality of sets. However, we also stressed that the *ordinal* property of the set $W$ of whole numbers is essential to the same process. It is this order property of $W$ which we shall examine now; especially in its relation to the base and place concepts in numeration.

Children ordinarily learn the decimal system words for the natural numbers, even before they understand the number concept, by counting. Perhaps that is the reason some people prefer to call the numbers $N = \{1, 2, 3, \cdots\}$ the *counting numbers*. There is further disagreement of terminology, in that $W = \{0, 1, 2, 3, \cdots\}$ is called the set of natural numbers by some mathematicians. This should not be alarming because their preference is usually readily apparent in what they say about the numbers—if you have been forewarned of the possibility.

In counting, in the sense of listing the natural numbers in order either verbally or in writing, there are two basic methods. The more common method is the memorized sequence, since it is learned first. This procedure, for the decimal system, is clearly illustrated by the odometer (mileage indicator) on the family car. Very few of us have ever seen one showing all zeros, but if we had such an opportunity, the odometer would appear as

$$000000$$

As the car moves forward, the dial at the right turns slowly until a 1 appears:

$$000001$$

No other dial moves during the time in which the right dial turns successively through the digits 2, 3, 4, 5, 6, 7, 8, and 9; but as it completes its cycle and 0 begins to show again, the next dial starts to move and advances to 1:

$$000010$$

The cycle is repeated, and some of the successive readings look like this:

$$000011$$
$$000012$$
$$\cdots$$
$$000019$$
$$000020$$
$$000021$$

The odometer is simply a mechanical device for counting. Such devices are used in many daily activities and we are hardly aware of them.

This illustration may seem pointless, since each of us knows how to count. However, while it is generally true that everyone knows how to count in base ten, how many people know how to count in base seven? Imagine dials, such as the odometer, with only the digits 0, 1, 2, 3, 4, 5, 6 on them, and you will see more clearly the derivation (and continuation) of the following counting: 1, 2, 3, 4, 5, 6, 10, 11, 12, 13, 14, 15, 16, 20, 21, 22, 23, 24, 25, $\cdots$.

*Numerals and Bases*

The second basic method of counting is essentially the method which was used to introduce the digits as symbols for cardinal numbers. Each succeeding number is found by adding one to the preceding number. This simple relation between two succeeding whole numbers has been used as a basis for a rigorous development (where everything is proved carefully) of the whole numbers, including the properties of addition and multiplication which we have not yet studied. An Italian mathematician, G. Peano, made one of the earliest and best known developments. In his system, he made only five basic assumptions. Although we shall attempt nothing so ambitious in our development, since we are working with those same numbers having those same properties, we will see a sample of his development in the next chapter.

*Exercise 3.6*

1. Write decimal numerals for the numbers represented by
   (a) $te_{twelve}$  (b) $21t_{twelve}$  (c) $4t6_{eleven}$

2. Write numerals in the following bases for the number one thousand nine hundred sixty-five.
   (a) base eight  (b) base eleven  (c) base twelve

3. Fill in the following blanks.
   (a) $415_{six}$ = _____ nine
   (b) $415_{nine}$ = _____ four
   (c) $2202_{three}$ = _____ twelve

*4. Count to twenty in the base three system using %, $, # as digits for zero, one and two respectively.

*5. Count to twenty in base four using the familiar digits 0, 1, 2, 3.

6. Is the sentence "$40_{six} = 60_{four}$" true, false, or meaningless? Explain.

# 4

# The Ordinal Property of $W$

The purpose of this chapter is to show how addition and multiplication of whole numbers can be developed from the ordinal property of the whole numbers. In Chapters 5 and 6, we will continue the development based upon the cardinal number concept introduced in Chapter 2. Since we have just seen, in the case of the odometer, that machine calculations are based upon this counting principle, we have reached an appropriate place for this development. We will also develop several properties essential to algorithmic processes, which we use to perform our numerical calculations and which will appear in later chapters.

## PEANO'S AXIOMS

The axioms which form the logical basis for this brief development were evolved by G. Peano (1858–1932), an Italian mathematician. Although they are usually worded to describe the ordered set of natural numbers $N = \{1, 2, 3, \cdots\}$, they apply equally well to the set $W = \{0, 1, 2, 3, \cdots\}$. The following set of axioms is a modification of those of Peano.

### Axiom 1

$0 \in W$.

### Axiom 2

For each $n \in W$, there is a unique $n' \in W$ called the *successor of* $n$; $0' = 1$.

## Axiom 3

If $m, n \in W$ such that $m' = n'$ then $m = n$.

## Axiom 4

There is no $m \in W$ for which $m' = 0$.

## Axiom 5

If $S \subseteq W$ such that (1) $0 \in S$, and (2) $n' \in S$ for every $n \in S$, then $S = W$.

As might be expected, the successor of an element of $W$ is simply the element which follows it in the natural ordering $W = \{0, 1, 2, 3, \cdots\}$. Thus $0' = 1, 1' = 2, 2' = 3$, etc. The significance of this concept is emphasized in the following definition of addition.

*Definition 4.1.* For every $m, n \in W$, (1) $n + 1 = n'$ and (2) $m + n' = (m + n)'$.

Using part (1) of the definition, part (2) may be written $m + (n + 1) = (m + n) + 1$. At first glance, this definition may seem inadequate. However, by Axiom 2, Definition 4.1 uniquely identifies $n + 1$ for every $n \in W$. For example: $0 + 1 = 0' = 1, 1 + 1 = 1' = 2, 2 + 1 = 2' = 3$; etc.

An elementary sum which is not quite so obvious is $m + 0$. Since this form is not covered by Definition 4.1, we shall have to use a longer and more devious approach.

$m + 0' = (m + 0)'$ by part (2) of Def. 4.1.
$m + 0' = m + 1$ since $0' = 1$, and
$m' = m + 1$ by part (1) of Def. 4.1.

Thus
$m' = (m + 0)'$ since $m' = m + 1 = m + 0'$ (substitution)
$m = m + 0$ by Axiom 3.

Thus we find that $m + 0 = m$. Let us formalize this discovery as a theorem for future reference:

## Theorem 4.1

For every $m \in W, m + 0 = m$.

As a second example of the use of Definition 4.1, we develop a few of the so-called addition facts. We already know that $1 + 0 = 1, 2 + 0 = 2, 3 + 0 = 3$, etc., by Theorem 4.1. We also have $0 + 1 = 1, 1 + 1 = 2, 2 + 1 = 3, 3 + 1 = 4$, etc., by direct application of Definition 4.1 to the recognition of the respective successors of 0, 1, 2, 3, etc. It appears that we can find the

result of adding either 0 or 1 to any whole number. The next step would seem to be an attempt to find a way to add 2 to a whole number. This general problem is left as an exercise but, as an illustration, we will show how to identify $5 + 2$. Now,
$$2 = 1'$$
so that
$$5 + 2 = 5 + 1'$$
while
$$5 + 1' = (5 + 1)'$$
by part (2) of Definition 4.1. Presumably, we have already found that $5 + 1 = 6$, so that we now have
$$5 + 2 = 6'$$
or
$$5 + 2 = 7$$
from the basic knowledge that $6' = 7$.

The following exercise will help you assimilate the basic concepts of Peano's method before proceeding further.

## Exercise 4.1

1. Write the numeral for $n' \in W$ if (a) $n = 10$, (b) $n = 47$, (c) $n = 1000$.
2. Write the numeral for $n \in W$ if (a) $n' = 476$, (b) $n' = 10,000$, (c) $n' = 0$.
3. From Peano's postulates, show that $W$ has no last element.
4. Use Definition 4.1 to find the element in $W$ defined by $3 + 2$.
*5. Prove that $m + 2 = (m')'$ for every $m \in W$.
6. From the knowledge that $5 + 2 = 7$, show how to identify $5 + 3$.
*7. Show that $2 + 2 = 3 + 1$ from Definition 4.1.
*8. Use Definition 4.1 and the result of Problem 7 to show that $2 + 3 = 3 + 2$.

## FINITE INDUCTION

From the preceding examples and exercises, it should be clear that the addition facts can be built up (rather slowly, to be sure) by performing *in order* addition by 0, 1, 2, 3, 4, etc. Can we be sure that the system will never break down? After all, we can only proceed from the sum $m + n'$. In a sense, Peano's fifth axiom was supplied to answer this question. First we shall analyze Axiom 5 (which we have not yet used) and then we shall illustrate its use. This axiom is often called the axiom of *finite induction*.

To analyze Axiom 5, consider a set $S$, about which you know nothing except that $S \subseteq W$, (1) $0 \in S$ and (2) for every $s \in S$, $s' \in S$ also. Now $0 \in S$

by (1); then by (2), $0' = 1 \in S$ also. Then, since $1 \in S$, $1' = 2 \in S$, also by part (2). But $2 \in S$ implies $2' = 3 \in S$. By now you should be assured that every element of $W$ is in $S$; i.e., that $W \subseteq S$. But if $W \subseteq S$ and $S \subseteq W$, then $S = W$. Of course we have no more assurance that this procedure will not break down than that the addition procedure previously discussed will not. This is the reason Peano labeled this assertion an axiom—an *assumption*. It should be noted that the situation used to state the axiom is probably as simple as could be found.

Now we turn our attention to the use of Axiom 5 in making proofs. The following theorem is the assurance that the sum $m + n$ can be identified in $W$ no matter what $m, n \in W$ are.

## Theorem 4.2

For every $m, n \in W$, Definition 4.1 uniquely defines $m + n \in W$.

*Proof*: Let $m$ be any element of $W$, and let $S$ be those elements $n \in W$ for which $m + n \in W$ is uniquely defined by Definition 4.1. Now $0 \in S$ since $m + 0 = m \in W$ uniquely by Theorem 4.1. This establishes (1) of Axiom 5. Now if $k \in S$, then $m + k \in W$ and is uniquely defined (by the definition of $S$). Then by Axiom 2, $(m + k)' \in W$ and is unique. But by Definition 4.1, $(m + k)' = m + k'$ so that $m + k'$ is a unique element of $W$, which means that $k' \in S$ also. This satisfies condition (2) of Axiom 5, so that $S = W$; i.e., for every whole number $n$, $m + n$ is a unique element of $W$.

## ADDITION PROPERTIES

We have just shown that the sum of two whole numbers is also a whole number. In general, when every pair of not necessarily distinct members of a set combine by an operation always to produce a member of the same set, that set is said to be *closed under that operation*. We have already proved that $W$ is closed under addition. This is not a particularly startling discovery, but it is a significant property of addition because it is not true of either subtraction or division.

In discussing some of the following properties, it will be convenient to use a symbol for a mathematical *system*. Briefly, a system is a set together with relations between its elements and operations on its elements. For example, $W$ with the *equals* and *greater than* relations and with addition constitutes a system. We will identify such a system by writing a set symbol followed by a semicolon, followed in turn by the operation symbols separated by commas, enclosing the entire composite symbol in parentheses. Thus $(W; +)$ represents the *system of whole numbers under addition*.

From our investigation of Peano addition in $(W; +)$ we know only how to add two whole numbers. What happens when $m, n, k \in W$, and we examine

$m + n + k$? Which two numbers should be added first? From prior experience, we would be inclined to say that it makes no difference. Certainly

$$(2 + 3) + 4 = 5 + 4 = 9$$

and

$$2 + (3 + 4) = 2 + 7 = 9$$

But if

$$(m + n) + k = m + (n + k)$$

for all whole numbers $m$, $n$ and $k$, it would be much more satisfying to have the fact verified by proof. This property of addition is called the *associative* property, and is verified in the following theorem.

## Theorem 4.3

Addition is associative in $(W; +)$.

*Proof*: Let $m$ and $n$ be any whole numbers and let $S \subseteq W$ for which $s \in S$ iff $(m + n) + s = m + (n + s)$. (1) $0 \in S$ because $(m + n) + 0 = m + n = m + (n + 0)$ by applying Theorem 4.1 twice. (2) Suppose $k \in S$. Then $k$ is a whole number for which $(m + n) + k = m + (n + k)$. Now examine $k'$: $(m + n) + k' = [(m + n) + k]'$ by part (2) of Definition 4.1, and $[(m + n) + k]' = [m + (n + k)]'$ since $(m + n) + k = m + (n + k)$. Then $[m + (n + k)]' = m + (n + k)' = m + (n + k')$ by applying part (2) of Definition 4.1 twice. Thus we have $(m + n) + k' = m + (n + k')$ which means that $k' \in S$ also. Since (1) and (2) of Axiom 5 are satisfied, it follows that $S = W$; or $(a + b) + c = a + (b + c)$ for whole numbers $a, b, c$.

Once again, this is an important property, since it is not true of subtraction. For example, $(11 - 6) - 4 = 5 - 4 = 1$ while $11 - (6 - 4) = 11 - 2 = 9$, so that $(11 - 6) - 4 \neq 11 - (6 - 4)$.

It is questionable whether all of the succeeding facts should be called theorems; but for the sake of convenience, we will do so. Although several results are obtained, the property of major importance is that addition is *commutative*; i.e., $m + n = n + m$ for every $m, n \in W$. You may recall that you were asked to show this for $2 + 3$ and $3 + 2$ in Problem 8 of Exercise 4.1.

## Theorem 4.4

$0 + m = m$ for every $m \in W$.

*Proof.* Let $S \subseteq W$ such that $s \in S$ iff $0 + s = s$. (1) $0 \in S$ since $0 + 0 = 0$ by Theorem 4.1. (2) If $k \in S$ then $0 + k = k$. Then $0 + k' = (0 + k)'$ by part (2) of Definition 4.1, which with $0 + k = k$ implies that $0 + k' = k'$. Thus if $k \in S$, then $k' \in S$, also. Since conditions (1) and (2) of Axiom 5 are satisfied, we have $S = W$, or $0 + m = m$ for every $m \in W$.

*Corollary.* $m + 0 = 0 + m$ for every $m \in W$.

Strictly speaking, a *corollary* is a statement implied directly and almost solely by the preceding theorem. This statement relies as heavily upon Theorem 4.1 as upon Theorem 4.4, since its proof depends upon both the facts that $m + 0 = m$ and $0 + m = m$ for every $m \in W$.

## Theorem 4.5

$1 + m = m'$ for every $m \in W$.

*Proof*: Let $S \subseteq W$ such that $s \in S$ iff $1 + s = s'$. (1) $0 \in S$ because $1 + 0 = 1 = 0'$ by Theorem 4.1 and Axiom 2. (2) If $k \in S$ then $1 + k = k'$. Also $1 + k' = (1 + k)' = (k')'$ by part (2) of Definition 4.1, and substituting $k'$ for $1 + k$. Thus, $k \in S$ implies $k' \in S$ also. By Axiom 5, $S = W$, or $1 + m = m'$ for every $m \in W$.

*Corollary.* $m + 1 = 1 + m$ for every $m \in W$.

The stage is now set for proving the theorem of commutativity of addition in the system $(W; +)$. The corollaries of the two preceding theorems will help greatly in the proof.

## Theorem 4.6

Addition is commutative in $(W; +)$.

*Proof*: Let $m$ be any whole number and let $S \subseteq W$ for which $s \in S$ iff $m + s = s + m$. (1) Then $0 \in S$ since $m + 0 = 0 + m$ by the corollary of Theorem 4.4. (2) If $k \in S$ then $m + k = k + m$. Then $m + k' = (m + k)' = (k + m)'$ by part (2) of Definition 4.1 and substituting $k + m$ for $m + k$. By Theorem 4.5, $(k + m)' = 1 + (k + m)$, and by associativity and another application of Theorem 4.5, $1 + (k + m) = (1 + k) + m = k' + m$. Thus $m + k' = k' + m$, so that $k' \in S$ also. By Axiom 5, then, $S = W$ so that $m + n = n + m$ for all $m, n \in W$.

Although our immediate interest is the collection of addition properties in $(W; +)$ which we have discussed, it is hoped that the examples of proof by induction (Axiom 5) will enable you to make the simple proofs called for in the exercises. Careful pronunciation of the words *commutative* and *associative* will help accurate spelling, and recognition of the common meanings of *commute* and *associate* will distinguish between the two concepts.

## Exercise 4.2

1. Which of the following sets are closed under addition?
    (a) $E = \{0, 2, 4, 6, \cdots\}$  (c) $D = \{1, 3, 5, 7, \cdots\}$
    (b) $T = \{0, 5, 6, 9, \cdots\}$

**2.** If $S$ represents the collection of all sets in a particular universe, then sets $A = \{a, b, c\}$ and $B = \{x, y\}$ are *elements* of $S$ while union and intersection are operations with sets. (a) Does $A \cup B = B \cup A$? (b) Does $A \cap B = B \cap A$?

**3.** Do you believe that $S \cup T = T \cup S$ and $S \cap T = T \cap S$ for all sets $S$ and $T$? What would you call this property?

**\*4.** If $A \cup B = B \cup A$, prove that $n(A) + n(B) = n(B) + n(A)$.

**5.** Demonstrate what is meant by the statement "Set intersection is associative." Is it true?

**6.** Combine numbers as indicated by the grouping symbols:

(a) $(3 + 2) + 4$      (c) $[(4 + 1) + 1] + 3$
(b) $(4 + 2) + 3$      (d) $(4 + 1) + (1 + 3)$

**7.** Name the property which justifies each of the following statements:

(a) $3 + 5 = 5 + 3$
(b) If $m$, $n$ are even numbers then $m + n$ is an even number; i.e., one of $\{0, 2, 4, 6, 8, \cdots\}$.
(c) $2 + (5 + 3) = (2 + 5) + 3$
(d) $(2 + 5) + 3 = 3 + (2 + 5)$
(e) $(2 + 5) + 3 = (5 + 2) + 3$

**8.** Apply the theorems of this section to show that $(2 + 5) + 3$ can also be written as $5 + (2 + 3)$. Show your work.

**\*9.** We have shown that $=$ is a *transitive* relation; i.e., if $a = b$ and $b = c$ then $a = c$. This statement involves three symbols: $a$, $b$ and $c$. By the finite induction axiom prove that if $a = a_1$, $a_1 = a_2$, $\cdots$, and $a_{n-1} = a_n$ then $a = a_n$ for every $n \in N$.

**\*10.** We have shown that 0 has the *zero* property; i.e., that $m + 0 = 0 + m = m$ for every $m \in W$. Use the properties we have developed to show that if there were *another* zero, symbolized by $z$, then $z = 0$. (Hint: What happens when you calculate $z + 0$ or $0 + z$?)

## MULTIPLICATION IN $W$

After a child has had experience in performing addition and subtraction, he is introduced to the concept of multiplication, usually through the avenue of repeated addition. This early association of addition with subtraction has certain pedagogical advantages which will become apparent in Chapter 7, when we study inverse operations. However, subtraction performed in the set $W$ of whole numbers has the logical disadvantage that it is not always defined. We made this same observation a little earlier when we stated that $W$ is not closed under subtraction. Since $W$ is closed under multiplication (to be proved shortly) and since we are attempting a logical development, we turn now to the problem of describing the results and properties of multiplication.

The definition used for Peano addition provided both a starting place (how to add 1) and a means of continuing. This enabled us to show how to add larger numbers, and also how to add zero. Although it would have deprived us of an excellent example of the use of Axiom 3 (see proof of Theorem 4.1), it would have been more correct from a logical viewpoint to have used Theorem 4.1 as part (1) of Definition 4.1. The definition chosen for multiplication uses this latter viewpoint.

*Definition 4.2.* For all $m, n \in W$, (1) $m \times 0 = 0$ (read $m$ *times* 0) and (2) $m \times n' = (m \times n) + m$.

For children, the case $m \times 0 = 0$ is one of the most difficult to accept on an intuitive basis. From the concept of repeated addition, it is easy to convince them that $1 \times 0 = 0$; $2 \times 0 = 0 + 0 = 0$; $3 \times 0 = 0 + 0 + 0 = 0$; etc. A simple induction proof establishes $m \times 0 = 0$ for every $m \in N$, but $0 \times 0$ or $0 \times 1$ cannot be treated so easily.

EXAMPLE: Let us illustrate the nature and adequacy of Definition 4.2 for the case $m = 3$. $3 \times 0 = 0$ by part (1). Then $3 \times 1 = 3 \times 0' = (3 \times 0) + 3$ by part (2), while $(3 \times 0) + 3 = 0 + 3 = 3$ by the previous multiplication facts and known addition facts; thus $3 \times 1 = 3$. $3 \times 2 = 3 \times 1' = (3 \times 1) + 3 = 3 + 3 = 6$. See if you can supply the reasons.

As in the case for addition, after establishing the facts for multiplication of zero [part (1) of Definition 4.2] the next consideration might well be the facts of the multiplication of 1, as we see in the following brief theorem.

# Theorem 4.7

$m \times 1 = m$ for every $m \in W$.

*Proof.* For every $m \in W$, $m \times 1 = m \times 0' = (m \times 0) + m$ by part (2) of Definition 4.2. Then, since $m \times 0 = 0$ by part (1) of Definition 4.2, $(m \times 0) + m = 0 + m = m$ by Theorem 4.4. Thus $m \times 1 = m$ for every $m \in W$.

We shall presently show that $1 \times m = m$ also. This should remind you of a similar property of the number zero under addition. In other words, the facts that
$$m + 0 = 0 + m = m$$
and
$$m \times 1 = 1 \times m = m$$
for every $m \in W$ are similar facts and the common property should be worthy of note.

*Definition 4.3.* If $*$ represents an operation between pairs of elements of a set $S$, then $e \in S$ is an *identity* for $*$ iff $e * s = s * e = s$ for every $s \in S$.

Thus, we say that $0 \in W$ is an identity for addition and $1 \in W$ is an identity for multiplication. We also use the expressions *additive identity* and *multiplicative identity*.

The following theorems establish properties for multiplication which are similar to those established for addition. The same descriptive words (commutative and associative) will be used. However, Theorem 4.8 introduces a new concept which we will illustrate first by example. If a salesman sold four 53-dollar vacuum cleaners one day and six the next, and waited until the end of the second day to figure his total sales for the two-day period, he would have two ways to perform the calculation. He could figure each day's sales separately and then add them; or he could first total the number of machines sold over the two-day period and then multiply that figure by 53. These calculations can be shown symbolically as

$$(1) \quad (4 \times 53) + (6 \times 53)$$

and

$$(2) \quad (4 + 6) \times 53$$

Both of these calculations turn out the same (530 dollars), illustrating the general property that

$$(a \times b) + (a \times c) = a \times (b + c)$$

or, in the form of (1) and (2),

$$(b \times a) + (c \times a) = (b + c) \times a$$

We describe this property by saying that *multiplication is distributive over addition*. Thus, if 7 is to be multiplied by the sum $10 + 3$ in the order

$$7 \times (10 + 3)$$

it would be permissible to *distribute* this multiplication by 7 *over* or among the numbers being added together, as in

$$(7 \times 10) + (7 \times 3)$$

Notice that this is what we normally do in multiplying 13 by 7; the 3 and the 10 are multiplied separately, and the results are added. This idea will be explored more extensively when the algorithms are analyzed in Chapters 5 and 6. Our interest at the moment is to establish this property for the system $(W; +, \times)$ involving the set of whole numbers and the operations of addition and subtraction.

## Theorem 4.8

Multiplication is distributive over addition in $(W; +, \times)$.

*Proof.* Let $m$ and $n$ be any elements of $W$ and let $S \subseteq W$ such that $s \in S$ iff $m \times (n + s) = (m \times n) + (m \times s)$. (1) $0 \in S$ because $m \times (n + 0) = m \times n$ and $(m \times n) + (m \times 0) =$

$(m \times n) + 0 = m \times n$ also, by the properties of zero already proved. This satisfies condition (1) of Axiom 5. (2) If $k \in S$ then $m \times (n + k) = (m \times n) + (m \times k)$. Then $m \times (n + k') = m \times (n + k)'$ by (2) of Definition 4.1, and $m \times (n + k)' = m \times (n + k) + m$ by (2) of Definition 4.2. But $m \times (n + k) + m = [(m \times n) + (m \times k)] + m$ since $k \in S$, while $[(m \times n) + (m \times k)] + m = (m \times n) + [(m \times k) + m]$ by Theorem 4.3 (associativity of addition). Finally, $(m \times k) + m = m \times k'$ by (2) of Definition 4.2, so that $(m \times n) + [(m \times k) + m] = (m \times n) + (m \times k')$, or $m \times (n + k') = (m \times n) + (m \times k')$. This shows that $k' \in S$ also, satisfying condition (2) of Axiom 5 so that $S = W$. Thus we have proved that $a \times (b + c) = (a \times b) + (a \times c)$ for all $a, b, c \in W$.

As you may expect from our earlier remarks, the next important theorem establishes the fact that multiplication in $(W; +, \times)$ is commutative. In order to keep its proof relatively clear and brief, we first establish two preliminary results.

## Theorem 4.9

$0 \times m = 0$ for every $m \in W$.

*Proof:* Let $S \subseteq W$ such that $s \in S$ iff $0 \times s = 0$. (1) $0 \in S$ because $0 \times 0 = 0$ by Definition 4.2. (2) If $k \in S$ then $0 \times k = 0$, so that $0 \times k' = 0 \times k + 0$ by part (2) of Definition 4.2, and $0 \times k + 0 = 0 + 0 = 0$ from the fact that $k \in S$ and Theorem 4.1. Thus $0 \times k' = 0$ so that $k' \in S$ also. Since (1) and (2) of Axiom 5 are satisfied, $S = W$, or $0 \times m = 0$ for every $m \in W$.

*Corollary.* $m \times 0 = 0 \times m$ for every $m \in W$.

## Theorem 4.10

$m' \times n = (m \times n) + n$ for all $m, n \in W$.

*Proof:* Let $m$ be any element of $W$ and let $S \subseteq W$ such that $s \in S$ iff $m' \times s = (m \times s) + s$. (1) $0 \in S$ because $m' \times 0 = 0$ and $(m \times 0) + 0 = 0 + 0 = 0$ by Definition 4.2 and Theorem 4.1, so that $m' \times 0 = (m \times 0) + 0$. (2) If $k \in S$ then $m' \times k = (m \times k) + k$; also $m' \times k' = (m' \times k) + m'$ by Definition 4.2, so that $(m' \times k) + m' = [(m \times k) + k] + m'$. But since addition is associative, $[(m \times k) + k] + m' = (m \times k) + (k + m')$, and by Definition 4.1 $(m \times k) + (k + m') = (m \times k) + (k + m)'$. Then since addition is commutative $(m \times k) + (k + m)' = (m \times k) + (m + k)'$, and $(m \times k) + (m + k)' = (m \times k) + (m + k')$ by Definition 4.1 again. Then by associativity of addition and Definition 4.2 in turn, $(m \times k) + (m + k') = [(m \times k) + m] + k' = (m \times k') + k'$. Thus we have shown that $m' \times k' = (m \times k') + k'$, so that $k' \in S$ also. Since conditions (1) and (2) of Axiom 5 are met, $S = W$, so that $m' \times n = (m \times n) + n$ for all $m, n \in W$.

This lengthy proof has the saving virtue of making the proof of our basic theorem short and easy. The theorem on commutativity of multiplication follows.

## *Theorem 4.11*

Multiplication is commutative in $(W; +, \times)$.

*Proof.* Let $m$ be any element of $W$ and let $S \subseteq W$ such that $s \in S$ iff $m \times s = s \times m$. (1) $0 \in S$ because $m \times 0 = 0 \times m$ by the corollary of Theorem 4.9. (2) If $k \in S$ then $m \times k = k \times m$. Now $m \times k' = (m \times k) + m$ by Definition 4.2, and since $k \in S$, $(m \times k) + m = (k \times m) + m$. But by Theorem 4.10, $(k \times m) + m = k' \times m$, so that $m \times k' = k' \times m$, or $k' \in S$ also. Since conditions (1) and (2) of Axiom 5 are met, $S = W$, or $m \times n = n \times m$ for all $m, n \in W$.

As a final example of the important results which follow so quickly from Peano's small list of axioms, we prove that $m \times n$ for $m \in N$ can legitimately be interpreted as *the sum of $m$ $n$'s*. This proof also hinges upon Theorem 4.10. Also, because of the logical difficulty inherent in the concept of *the sum of zero $n$'s*, induction will be on $N$ instead of $W$; i.e., $S \subseteq N$.

## *Theorem 4.12*

$m \times n$ is the sum of $m$ $n$'s for every $m \in N$ and $n \in W$.

*Proof:* Let $n$ be any element of $W$ and let $S \subseteq N$ such that $s \in S$ iff $s \times n$ is the sum of $s$ $n$'s. (1) $1 \in S$ because $1 \times n = n \times 1 = n$ by Theorems 4.11 and 4.7. (2) If $k \in S$ then $k \times n$ is the sum of $k$ $n$'s. Then $k' \times n = (k \times n) + n$ by Theorem 4.10, which is the sum of $k + 1$ or $k'$ $n$'s. Thus $k' \in S$ also, establishing both conditions of Axiom 5 (modified) so that $S = N$, or $m \times n$ is the sum of $m$ $n$'s for every $m \in N$ and $n \in W$.

We have now finished the brief formal development of the fundamental properties of the arithmetic of whole numbers by using the ordinal property of $W$. Since this is not the method used to introduce elementary school children to these ideas, we will derive them in Chapters 5 and 6 from the cardinal number concept, and use them there to develop the addition and multiplication procedures. In this chapter, however, you have seen some simple illustrations of the use of the very powerful induction principle which has been an aid in the proof of many important mathematical theorems. You should have some appreciation for the intuitive simplicity of Peano's axioms and the natural definitions for addition and multiplication. For example, everyone who knows how to multiply should realize that $6 \times 18$ is 6 more than $6 \times 17$, or

$$6 \times 18 = (6 \times 17) + 6$$

64                Elementary Mathematical Structure

which is simply the application of Definition 4.2. The following exercise will give you an opportunity to test your grasp of the concepts in this section.

## Exercise 4.3

1. Which of the following sets are closed under multiplication?
   (a) $E = \{0, 2, 4, 6, \cdots\}$   (c) $D = \{1, 3, 5, 7, \cdots\}$
   (b) $T = \{0, 3, 6, 9, \cdots\}$

*2. Let $A = \{a, b\}$, $B = \{b, c, d\}$ and $C = \{c, d, e\}$. Examine the following statements to arrive at a verdict of true or false.
   (a) $A \cap (B \cup C) = (A \cap B) \cup (A \cap C)$
   (b) $A \cup (B \cap C) = (A \cup B) \cap (A \cup C)$

*3. What is meant by the statement "Set intersection is distributive over set union?" Is it true?

4. In the following pairs of numbers, identify the numbers by combining numbers as indicated by the grouping symbols, then indicate whether or not the two numerals in each pair are equal.
   (a) $3 \times (4 + 7)$              $(3 \times 4) + 7$
   (b) $(3 + 4) \times 7$              $3 + (4 \times 7)$
   (c) $(3 \times 4) + (3 \times 7)$   $3 \times (4 + 7)$
   (d) $2 \times (3 + 1) \times 4$     $(2 \times 3) + (1 \times 4)$
   (e) $2 \times (3 \times 1) \times 4$ $(2 \times 3) \times (1 \times 4)$

5. Justify each of the following statements by a definition or by the name of a property:
   (a) $(3 + 7) + 1 = 3 + (7 + 1)$
   (b) $12 \times 5 = (11 \times 5) + 5$
   (c) $(3 + 4) \times 5 = 5 \times (3 + 4)$
   (d) $5 \times (3 + 4) = 15 + 20$
   (e) $(5 \times 3) \times 4 = 5 \times 12$

6. From Definition 4.2 and Theorem 4.7, we know that $3 \times 0 = 0$ and $3 \times 1 = 3$. Apply Definition 4.2 to calculate:
   (a) $3 \times 2$    (b) $3 \times 3$

7. Have we developed a means of calculating $3 \times 5$ from the results of Problem 6 without first having to calculate $3 \times 4$? Demonstrate.

*8. Show that if $e$ is an identity for multiplication in $(W; +, \times)$ then $e = 1$.

9. Prove that $1 \times m = m$ for every $m \in W$. You may use any material previous to Theorem 4.10.

*10. Prove by induction that
$$a \times (b_0 + b_1 + \cdots + b_n) = (a \times b_0) + (a \times b_1) + \cdots + (a \times b_n)$$
for every $n \in W$.

# 5

# Addition of Whole Numbers

Now that we have examined the systems $(W; +)$ and $(W; +, \times)$ from a rather technical and abstract viewpoint in Chapter 4, we shall return to the more intuitive approach used in teaching children these concepts. In this chapter we will briefly explore the concept of a mathematical system, with the system $(W; +)$ as our primary example. After examining a mechanical device for counting which employs the place value concept, we will develop the written algorithm, or procedure, using the addition properties of the system $(W; +)$. Finally, we will discuss the concept of linear measure, and use it to develop the concept of a number line, a device which vies in popularity with the set concept in the teaching of arithmetic. We begin our investigation with a brief discussion of the somewhat old-fashioned concept of a mathematical operation.

## MATHEMATICAL OPERATIONS

Union and intersection are often referred to as *operations* on or with sets. One objection to this application of the word operation is that it has the connotation of *doing something* with the sets. Although we speak of building sets, it is probably better logic to think in terms of the *existence* of sets. For example, the elements of the set $A = \{a, b, c\}$ existed before it even occurred to us to associate them as a set. Similarly, if we think of the set $B = \{b, c, d, e\}$, then the set $\{a, b, c, d, e\}$ certainly exists without our uniting the elements of $A$ and $B$ to form it. In effect, we merely recognize its existence and its relation to sets $A$ and $B$, and invent symbols ∪ and = to show the relation

$$A \cup B = \{a, b, c, d, e\}$$

which we recognize.

It should be clear that the identification of sets $A$ and $B$ alone is not sufficient to identify the set $\{a, b, c, d, e\}$ because, for example, they identify the sets $\{b, c\}$ and $\{a, d, e\}$ equally well. To demonstrate further the need for defining an operation, we recall that certain corresponding cardinal numbers are associated with $A$ and $B$. For example, each of the numbers 7, 5, 2, 3 and 1 has a meaningful association with this pair of sets.

Now that we have introduced all of these confusing possibilities, let us examine again the nature of previous definitions which enable us to avoid such confusions. Suppose we had not made clear the way in which the word *or* was to be interpreted, and suppose we had said that "$A \cup B$ is the set composed of all elements belonging to $A$ or to $B$." Under these conditions, you would probably agree that $A \cup B$ might be any one of the sets

$$\{a, b, c\} \quad \{b, c, d, e\} \quad \{a, b, c, d, e\}$$

In such a case, we say that the operation *union* has not been well defined. Don't you agree? This possibility of ambiguity illustrates the desirability and the significance of the existence of *unique* successors, sums, and product, as discussed in Chapter 4. Let us clarify some of these concepts using several definitions.

*Definition 5.1.* A finite *sequence* from a set $S$ is an ordered list $(s_1, s_2, \cdots, s_n)$ of elements of $S$, which are not necessarily distinct. $n$ is the *length* of the sequence.

A sequence, as we have defined it, is usually called an *ordered n-tuple* (or one of several similar terms) by mathematicians. Since our immediate application of the idea differs somewhat from the standard context for $n$-tuples, the word *sequence* has been appropriated to convey the desired concept.

EXAMPLE: If $S = N$, then the elements of the standard counting sets $N_1 = \{1\}$, $N_2 = \{1, 2\}$, $N_3 = \{1, 2, 3\}$, etc., form sequences. Since a sequence allows duplication or repetition of elements, such lists as

$$(a, b, a, c) \text{ and } (a, c, b, a)$$

are also sequences. Although the last two sequences list the same elements the same number of times, they are distinct sequences because *the ordering is not the same.*

*Definition 5.2.* An *operation* on the elements of a set $S$ is a correspondence between certain sequences of $S$ and elements of a set $T$. The operation is *well defined* iff for each of the sequences there is exactly one element of $T$ which corresponds to it. It is a *binary* operation iff each sequence is of length two.

It should be recognized that this definition of operation is quite broad. Mathematicians have learned that a definition designed for some particular situation may have to be altered again and again for somewhat similar situations, whereas less restrictive wording can result in a classification to fit all the situations. We want Definition 5.2 to be descriptive of the familiar arithmetic operations addition, subtraction, multiplication and division, and it should also be broad enough to include operations involving sets, such as union, intersection, subtraction and complementation. In formulating this definition we have been careful to allow (1) repetition of elements and (2) an unspecified number of elements, but to require (3) a specified order to emphasize the commutative property. As you might suspect, there are situations in which it is not feasible to specify even the restriction (3). Although we will be concerned largely with binary operations involving ordered pairs of elements, the following examples illustrate the general nature of the concept of mathematical operation.

EXAMPLE: Addition as defined in this text is a well-defined binary operation on $W$. That is, for each sequence $(a, b)$ of length two from $S = W$, there is a unique element $a + b$ from $T = W$.

EXAMPLE: Subtraction is a well-defined binary operation on $W$. For certain ordered pairs of elements of $W$ there are correspondingly unique elements of $W$. For example, $(7, 3)$ corresponds to 4 in $W$, $(6, 6)$ to 0 in $W$, etc. It should be noted that more than one sequence from $S$ may correspond to the same element of $T$, as, for example, $(10, 6)$ also corresponds to 4 in $W$. Although there is no correspondent in $W$ for $(3, 7)$, we shall soon recognize a set $T$ such that every sequence of length two from $W$ will have a correspondent in $T$ for the operation of subtraction.

EXAMPLE: Let $S$ be the set of all subsets of $\{a, b, c, d, e\} = F$, with $A = \{a, b\}$, $B = \{b, c\}$, $C = \{c, d\}$ and $D = \{d, e\}$. If it were known that there is some operation on $S$ for which

$(A, C, B)$ corresponds to $\{a, b, c, d\}$

$(A, B, D)$ corresponds to $\{a, b, c, d, e\}$

and

$(B, C)$ corresponds to $\{b, c, d\}$

we might suspect that the operation is something like union, except that it is not necessarily binary in nature. This illustrates the more general concept of a union of sets as the set consisting of all of their elements without repetition and without regard for order.

EXAMPLE: Most of us have seen a set like $T = \{1, 4, 9, 16, 25, \cdots\}$ on certain types of aptitude tests, where the student is asked to write the next few numbers. In this case, each number is associated with a single element of $S$: 1 with 1; 2 with 4; 3 with 9; 4 with 16; 5 with 25; 6 with 36; and so on. This operation of finding the square of a number is an example of a *unary* operation, an operation involving *one* element.

These examples should make it clear that an operation need not be binary. It is both convenient and sufficient in most elementary mathematical situations, however, to choose to define the basic operations as binary operations. Imagine how much work would be involved in memorizing sums of *three* one-digit numbers! Since all of the operations discussed in this book are well-defined binary operations, we will not emphasize these properties to any great degree hereafter.

## OPERATION NOTATION

To denote a binary operation involving two elements of a set, it is customary to write symbols for the elements, with a symbol for the operation placed between them. For example, in the notation $A \cup B$, $A$ and $B$ represent the elements and $\cup$ represents the operation, whereas $A \cup B$ represents the element associated with the pair $A$ and $B$. In a more familiar situation, $3 + 4$ represents a number associated with the pair of numbers *three* and *four* by addition, which number is readily identified as *seven* but symbolized as surely by $3 + 4$ as by the numeral 7.

*Definition 5.3.* The symbol $a * b$ represents the unique element of set $T$ (of Definition 5.1) which is associated with the ordered pair $(a, b)$ of elements of set $S$ by the binary operation symbolized by $*$.

Thus for sets $A$ and $B$, the symbols $A \cup B$, $A \cap B$, and $A - B$ also represent sets, while $n(A) + n(B)$ and $n(A) - n(B)$ represent numbers. In all these cases, the symbols $\cup$, $\cap$, $-$, and $+$ represent operations.

## MATHEMATICAL SYSTEMS

The foregoing discussion has related to ideas of a basic set $S$, a basic operation on members of $S$, a secondary set $T$ whose elements are related to elements of $S$ by the operation, and a basic relation such as *equals* between pairs of elements of $S$ and of $T$. These are the basic ingredients of a *mathematical system*, although a system may be either slightly simpler or much more complex, depending on the number of operations and relations involved in it. We introduce a symbol for a system in which the symbol for the basic set $S$ is followed by a semicolon and a list of the operations, all enclosed in parentheses.

*Definition 5.4.* $(S; *)$ represents a mathematical *system* involving the elements of a set $S$ *under* a binary operation $*$. Specifically, $(W; +)$ represents the *system of whole numbers under addition*.

Each mathematical system has, besides the basic characteristics just described, its own peculiar properties which are determined by the nature of its elements, operations and relations. Some of these properties are very useful and are common to many systems, so mathematicians have identified and named them. We shall examine a few of these.

*Definition 5.5.* A set $S$ is *closed* under an operation $*$ if and only if $m * n \in S$ for all $m, n \in S$.

In other words, if $T \subseteq S$ (using $S$ and $T$ as in Definition 5.2) then $S$ is closed under the operation which defines $T$. If the elements of $S$ are the subsets of $A = \{a, b\}$ then $S$ is closed under the operation $\cup$ because the union of two subsets of $A$ is always a subset of $A$, hence a member of $S$. The fact that the union of two sets is always a set made it possible to use set union to define addition in $(W; +)$ and assured the existence of $m + n$ as an element of $W$ for each pair $m, n \in W$. We can restate Theorem 2.1 using the terminology of Definition 5.4.

# Theorem 5.1

The set $W$ is closed under addition.

In our discussion of set union as a binary or non-binary operation, we observed that our physical and mental limitations often make it necessary to examine objects and symbols in order, especially if we exhibit them in writing. We have seen that order is important in some cases, such as in the use of the elements of $N$ in counting, and is unimportant in others. For example, since the order in which the elements of a set are listed has no effect on the identity of the set, we would expect that the sets $A \cup B$ and $B \cup A$ would consist of the same elements no matter what sets $A$ and $B$ represent. That is,

$$A \cup B = B \cup A$$

for all sets $A$ and $B$. This is a significant property since all operations do not have it. If, for example, $A = \{a, b, c\}$ and $B = \{b, d\}$ then $A - B \neq B - A$ since $A - B = \{a, c\}$ while $B - A = \{d\}$.

*Definition 5.6.* A binary operation $*$ is *commutative* on a set $S$ if and only if $a * b = b * a$ for all $a, b \in S$.

The following theorem is stated without proof. It is left as an exercise for the student to argue its validity.

## Theorem 5.2

Addition is commutative in $(W; +)$.

Another property which has been recognized briefly in our discussion is demonstrated when three elements are to be combined by using a binary operation. If $a$, $b$ and $c$ are elements of a set $S$ with a binary operation $*$ and are to be written in order

$$a * b * c$$

then either $a * b$ is to be recognized as some $x \in S$ with the result

$$a * b * c = x * c$$

or $b * c = y$ with $y \in S$ so that

$$a * b * c = a * y$$

Of course, an arbitrary choice could be made by which the procedure could be defined. In many systems it is not necessary to make such a choice because both procedures yield the same result; that is,

$$x * c = a * y$$

This idea is more clearly presented by using grouping or punctuation symbols to write

$$(a * b) * c = a * (b * c)$$

where $a * b$ and $b * c$ being enclosed in parentheses indicates the need to identify each of these as elements of $S$ before performing the remaining operations.

*Definition 5.7.* A binary operation $*$ is said to be *associative* on a set $S$ if and only if $(a * b) * c = a * (b * c)$ for all $a, b, c \in S$.

We have already indicated that the union of an arbitrary number of sets consists of the same elements regardless of the manner in which they are initially grouped or ordered. In a more rigorous discussion of set theory, however, it is dutifully argued that for arbitrary sets $A$, $B$ and $C$, $(A \cup B) \cup C \subseteq A \cup (B \cup C)$ and also $A \cup (B \cup C) \subseteq (A \cup B) \cup C$ which proves that $(A \cup B) \cup C = A \cup (B \cup C)$. Thus, set union is associative, so for disjoint sets $A$, $B$ and $C$:

$$n[(A \cup B) \cup C] = n[A \cup (B \cup C)]$$

or

$$n(A \cup B) + n(C) = n(A) + n(B \cup C)$$

or

$$[n(A) + n(B)] + n(C) = n(A) + [n(B) + n(C)]$$

## Theorem 5.3

Addition is associative in $(W; +)$.

It must be emphasized that although $a$, $b$ and $c$ may represent distinct elements in Definitions 5.6 and 5.7, it is not necessary that they do so. In other words, if $*$ is an associative operation on $S$ and $b \in S$ then $b * b = b * b$ and $(b * b) * b = b * (b * b)$, for example. For an associative operation it is common practice to omit the grouping symbols; that is, to recognize that

$$a * b * c$$

is *defined* whether it is interpreted as

$$(a * b) * c$$

or as

$$a * (b * c)$$

## Exercise 5.1

1. Let $*b$ represent a *unary* operation on a *single* element $b$ of set $S$ which produces an element of $T$ twice as large as $b$. For the given set $S$ show the corresponding set $T$.
   (a) $S = \{0, 5, 10\}$ (b) $S = N$ (c) $S = \{\text{nickel, pint}\}$

2. Let $*(a, b, c)$ represent a *ternary* operation on a set $S$ of numbers which produces the same number as $(a \times b) + c$. For example, $*(2, 3, 1) = (2 \times 3) + 1 = 6 + 1 = 7$. If $S = \{0, 1, 2\}$, show the corresponding set $T$.

3. Let $*$ represent "clock addition" (on a twelve hour clock). Thus $9 * 7 = 4$ indicates that 7 hours after 9 o'clock it is 4 o'clock. Which of the following sets are closed under $*$?
   (a) $\{3, 6, 9, 12\}$ (d) $\{2, 7, 12\}$
   (b) $\{4, 8, 12\}$ (e) $\{2, 4, 6, 8, 10, 12\}$
   (c) $\{6, 12\}$

4. Let $S = \{A, B, C, \emptyset\}$ where $A = \{a\}$, $B = \{b\}$, $C = \{a, b\}$ and $\emptyset = \{\ \}$.
   (a) Is $S$ closed under set union?
   (b) Is $S$ closed under set intersection?
   (c) Is $S$ closed under set subtraction?

5. Using $S = \{A, B, C, \emptyset\}$ as in Problem 4, is set subtraction (a) commutative or (b) associative? Illustrate.

*6. Which of the following subsets of $W$ are closed under the ordinary addition of $(W; +)$?
   (a) $\{1, 3, 5, 7, \cdots\}$ (d) $\{0, 1\}$
   (b) $\{0, 5, 10, 15, \cdots\}$ (e) $\{8, 10, 12, 14, 16, \cdots\}$
   (c) $\{0\}$

*7. Use Definition 2.6 to argue that addition is commutative in $(W; +)$. Illustrate with an example showing $2 + 3 = 3 + 2$.

8. Combine numbers as indicated by the grouping symbols:
   (a) $(3 + 2) + 4$
   (b) $(4 + 2) + 3$
   (c) $(4 + 1) + (1 + 3)$
   (d) $[(4 + 1) + 1] + 3$
   (e) $7 + (0 + 5)$

9. Name the property which justifies each of the following statements:
   (a) $(3 + 5) + 7$ is a whole number.
   (b) $(3 + 5) + 7 = 7 + (3 + 5)$.
   (c) $7 + (3 + 5) = (7 + 3) + 5$.

10. Apply the principles of this section to show that $(2 + 5) + 3$ can also be written as $5 + (2 + 3)$. Show your work in a manner similar to parts (b) and (c) of Problem 9.

## COMPUTING DEVICES

Although it took thousands of years for man to evolve an efficient numeration system, it took even longer to devise written procedures (called *algorisms* or *algorithms*) for performing addition and the other arithmetic operations. In the meantime a variety of computing devices were invented for use in business transactions. Most of these machines were simply counting devices, ranging from counting boards used by merchants (which gave the name *counter* to the table across which merchandise is sold) to the *abacus*.

The abacus is an instrument of oriental origin which is still used by the Japanese and Chinese to make computations skillfully and quickly. It is usually a wooden frame holding several uniformly spaced lengths of wire upon which are strings of wooden beads. Each wire represents a place value, and the number of beads displaced upon it represents the digit for that place. In Figure 5.1, the numeral 2,035 is recorded on a simplified version of an

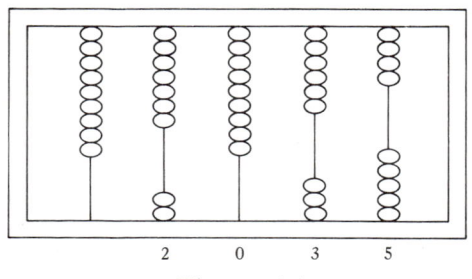

*Figure 5.1*

abacus. Notice that nine beads on each wire is sufficient, since the number represented by ten beads on a wire can also be represented by one bead on

*Addition of Whole Numbers* 73

the next wire to the left. A sophisticated version of the type commonly used by professionals is shown in Figure 5.2. It makes use of the intermediate

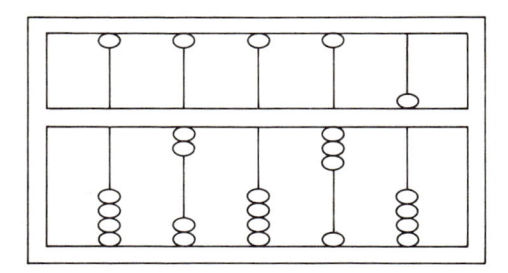

*Figure 5.2. Abacus.*

values introduced by the Romans in their numerals. Each top bead is equivalent to five immediately below it. The beads being used to represent the number are grouped next to the middle section of the frame. Again the number 2,035 is represented.

It should be recognized that the beads on the abacus serve as elements of a set. As the set elements on the first wire accumulate to ten, they are grouped into a subset which is then represented by a bead on the second wire. Ten of these beads is represented by a bead on the third wire, etc.

The procedure for adding on the abacus is best illustrated by comparing the collecting of the beads with the union of two sets. We will take an example using our definition of addition in terms of sets, show how this can be done on the abacus (simple form), and demonstrate a corresponding procedure using numerals.

EXAMPLE: Let us find the sum 23 + 45. First we represent these numbers by sets, using groups of tens, as in Figure 5.3. By definition, $n(A) + n(B) = n(A \cup B)$. Then, since $A \cup B$ consists of 2 + 4 = 6 tens and 3 + 5 = 8 ones, $n(A \cup B) = 68$.

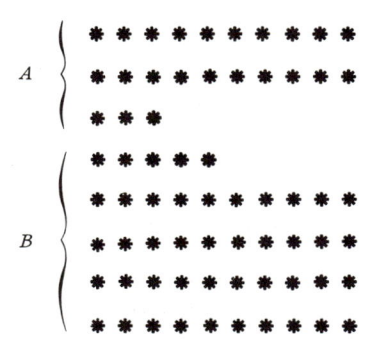

*Figure 5.3. Sets Representing 23 and 45.*

To perform the computation on the abacus, we first post one of the numbers, say 23, after which the first two wires look like those pictured in Figure 5.4.

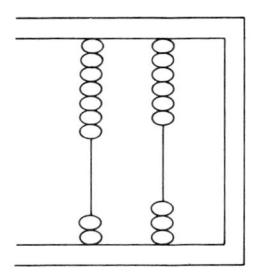

**Figure 5.4.   The Number 23.**

Then we displace 5 more beads on the first wire and 4 more on the second, so that the number 23 + 45 is shown on the abacus in Figure 5.5.

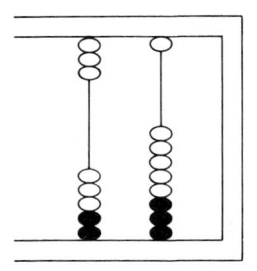

**Figure 5.5.   23 + 45 = 68.**

Everything which has been done with sets $A$ and $B$ and with the abacus can be done in writing, using the expanded forms of the numerals. Thus $23 = 2$ tens $+ 3$ and $45 = 4$ tens $+ 5$, so that

$$23 + 45 = (2 \text{ tens} + 3) + (4 \text{ tens} + 5)$$
$$= (2 \text{ tens} + 4 \text{ tens}) + (3 + 5)$$

making free use of the commutative and associative properties of addition. But

$$2 \text{ tens} + 4 \text{ tens} = (2 + 4) \text{ tens}$$

so that

$$23 + 45 = 6 \text{ tens} + 8 = 68$$

Of course, this procedure is laborious also, but it does demonstrate why the digits 2 and 4 are combined, while independently the units digits 3 and 5 are also combined. In the next sections we shall do similar work in the non-decimal

*Addition of Whole Numbers*

bases and graduate to the more convenient form

$$\begin{array}{r} 23 \\ 45 \\ \hline 68 \end{array}$$

## ADDITION TABLES

In the previous example, the sums $2 + 4$ and $3 + 5$ could be obtained by counting groups and elements of sets, or the beads on the abacus. When solving such a problem without these aids, it becomes necessary to have some other source for such basic knowledge. Furthermore, if we expect to develop skill and speed in the procedure of addition, we must be able to find the facts we need very quickly. One way to do this would be to record the facts in writing in some organized fashion, producing an *addition table*. The only facts which we need in such a table are the sums of pairs of digits. Such facts are often called *basic addition facts*. Since most of us spent hours memorizing these facts for base ten, we shall use other bases in our analysis here, so that we are not tempted to follow a procedure mechanically, without thinking about what we are doing.

EXAMPLE: In order to expand your experience to include work in several bases, we shall perform the work of these examples in base six. First we construct an addition table, by means of either of the definitions for addi-

*Table 12.*

| + | 0 | 1 | 2 | 3 | 4 | 5 |
|---|---|---|---|---|---|---|
| 0 | 0 | 1 | 2 | 3 | 4 | 5 |
| 1 | 1 | 2 | 3 | 4 | 5 | 10 |
| 2 | 2 | 3 | 4 | 5 | 10 | 11 |
| 3 | 3 | 4 | 5 | 10 | 11 | 12 |
| 4 | 4 | 5 | 10 | 11 | 12 | 13 |
| 5 | 5 | 10 | 11 | 12 | 13 | 14 |

tion. Let us find a numeral to represent the sum $312_{six} + 123_{six}$. First, we express the numerals in expanded form, to emphasize the principle of combining digits of the same place value. Thus,

$$312_{six} + 123_{six} = (3 \text{ thirty-sixes} + 1 \text{ six} + 2 \text{ ones})$$
$$+ (1 \text{ thirty-six} + 2 \text{ sixes} + 3 \text{ ones})$$
$$= (3 \text{ thirty-sixes} + 1 \text{ thirty-six})$$
$$+ (1 \text{ six} + 2 \text{ sixes}) + (2 \text{ ones} + 3 \text{ ones})$$
$$= (3 + 1) \text{ thirty-sixes} + (1 + 2) \text{ sixes}$$
$$+ (2 + 3) \text{ ones}$$
$$= 4 \text{ thirty-sixes} + 3 \text{ sixes} + 5 \text{ ones}$$
$$= 435_{six}$$

Now that the need for the associative and commutative properties of addition has been explained by the horizontal form, we will use the following more convenient form in our future work. This form combines the horizontal and vertical forms of addition.

$$\begin{array}{r} 3 \text{ thirty-sixes} + 1 \text{ six} \ \ + 2 \text{ ones} \\ + \ 1 \text{ thirty-six} \ \ + 2 \text{ sixes} + 3 \text{ ones} \\ \hline 4 \text{ thirty-sixes} + 3 \text{ sixes} + 5 \text{ ones} \\ = 435_{six} \end{array}$$

where the sums $3 + 1$, $1 + 2$ and $2 + 3$ are read directly from Table 12.

EXAMPLE: Our second example will involve the technique of *regrouping* or *carrying*. The need for this process occurs when the sum of digits is so large that it must be expressed as a two-digit numeral. For example, in base six, $4 + 5 = 13$. If these digits had represented the numbers 4 ones + 5 ones = 13 ones then the sum would need to be written as 10 ones + 3 ones (hence the term regrouping), which can be written 1 six + 3 ones. If 4 and 5 had been in sixes place, then 4 sixes + 5 sixes = 13 sixes = 10 sixes + 3 sixes = 1 thirty-six + 3 sixes. Notice how the place value system automatically forces the 1 into its correct place when the 3 occupies its place.

As our second example, we shall find a numeral in base six for $214_{six} + 125_{six}$.

$$\begin{array}{r} 2 \text{ thirty-sixes} + 1 \text{ six} \ \ + \ 4 \text{ ones} \\ + \ 1 \text{ thirty-six} \ \ + 2 \text{ sixes} + \ 5 \text{ ones} \\ \hline 3 \text{ thirty-sixes} + 3 \text{ sixes} + 13 \text{ ones} \\ = 3 \text{ thirty-sixes} + 3 \text{ sixes} + \ 1 \text{ six} + 3 \text{ ones} \\ = 3 \text{ thirty-sixes} + 4 \text{ sixes} + \ 3 \text{ ones} \\ = 343_{six} \end{array}$$

*Addition of Whole Numbers*

Since we observed that the digit 1 in 13 ones is forced into six's place when 3 is in ones place, let us see what happens when we do not write the place values, but merely observe them according to the relative positions in which we write the digits. We will, however, record the sum of each pair of digits separately in the proper places, supplying zeros in the vacant places to their right. Thus instead of writing 3 thirty-sixes, we shall write $300_{six}$.

$$\begin{array}{r} 214 \\ 125_{six} \\ \hline \left. \begin{array}{r} 13 \\ 30 \\ 300 \end{array} \right\} \text{Partial sums} \\ \hline 343_{six} \end{array}$$

These *partial sums*, 13, 30, and 300, are helpful in explaining this procedure to children, but as they gain skill and confidence in their addition facts we encourage them to write the following short form and, eventually, have them mentally carry the 1 of 13 to be combined with the other two sixes place digits.

$$\begin{array}{r} 1 \\ 214 \\ 125_{six} \\ \hline 343_{six} \end{array}$$

Of course, to add columns of several numerals with any facility at all requires the ability to remember a partial sum and combine it with the next digit. Such skill is developed through hours of practice in each grade, with problems of increasing length and difficulty. However, if you do not have this ability to a reasonable degree, you should practice consciously to improve your ability.

EXAMPLE: Here is one final example, to let you check your understanding of this algorithm:

$$453_{six} + 345_{six}$$

$$\begin{array}{r} 4 \text{ thirty-sixes} + 5 \text{ sixes} + 3 \text{ ones} \\ 3 \text{ thirty-sixes} + 4 \text{ sixes} + 5 \text{ ones} \\ \hline 11 \text{ thirty-sixes} + 13 \text{ sixes} + 12 \text{ ones} \end{array}$$

$= 1 \text{ two hundred sixteen} + 1 \text{ thirty-six}$
$\qquad\qquad\qquad\quad + 1 \text{ thirty-six} + 3 \text{ sixes}$
$\qquad\qquad\qquad\qquad\qquad\qquad + 1 \text{ six} + 2 \text{ ones}$

$= 1 \text{ two hundred sixteen} + 2 \text{ thirty-sixes} + 4 \text{ sixes} + 2 \text{ ones}$

$= 1242_{six}$

With partial sums:

$$453$$
$$345_{six}$$
---
$$12$$
$$130$$
$$1100$$
---
$$1242_{six}$$

Finally, with carrying:

$$543$$
$$345_{six}$$
---
$$1242_{six}$$

*Exercise 5.2*

1. What number is posted on each abacus shown?

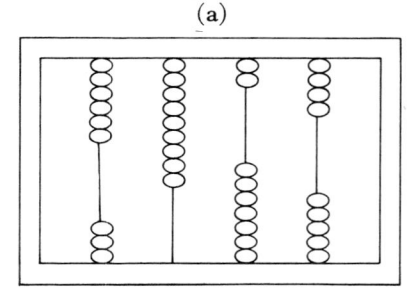

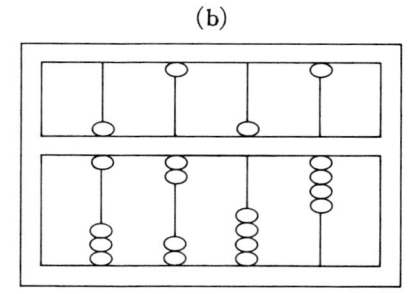

   (a)                               (b)

2. Draw pictures and post the number 7,604 on an abacus of each type as shown in parts (a) and (b) of Problem 1.

3. Suppose the picture shows a simple abacus for a non-decimal base. (a) What base does it use? (b) In that base, what numeral is posted on it? (c) What number is posted on it?

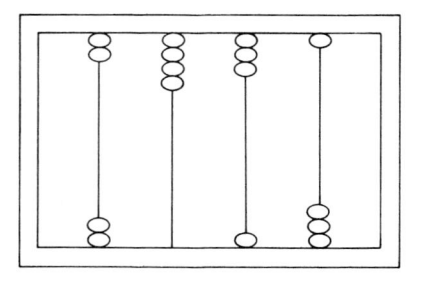

*Addition of Whole Numbers* 79

4. (a) Draw a simple abacus for a base four system and post the number one hundred. (b) Show with numerals the addition problem and solution illustrated in the picture.

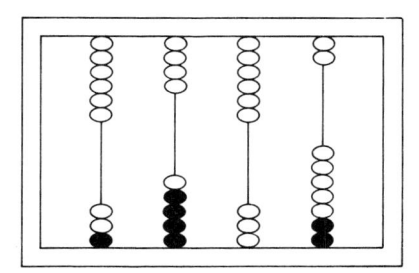

5. Use expanded form and Table 12 to show the addition $210_{six} + 135_{six}$.

6. Write the base-six expanded-form numeral 32 thirty-sixes + 15 sixes + 44 ones in standard base six (a) expanded; (b) digit form.

7. Perform the following additions using partial sums in base six (See Table 12).
   (a) 32
       13$_{six}$
   
   (b) 304
       123$_{six}$
   
   (c) 303
       253$_{six}$

8. Perform the additions of Problem 7 using carrying.

9. Make an addition table for the base five system with digits 0, 1, 2, 3, 4.

10. Perform the following additions in base five, using the table of Problem 9.
    (a) 132
        212$_{five}$
    
    (b) 333
        413$_{five}$
    
    (c) Two hundred six plus one hundred seventeen.

## GEOMETRY AND NUMBERS

One of the devices used as an aid to understanding of numbers and basic operations with numbers is the so-called *number line*, an application and extension of the concept of a ruler. Although everyone knows how to measure, this basic concept includes some situations which are as intricate and involved as the study of cardinality of sets. For this reason, among others, we shall digress briefly to lay a foundation in geometry which will help in the conceptual development of more sophisticated number systems.

It is significant, in the historical development of mathematics, that the method of analytic geometry made possible the development of the calculus which is probably the most powerful and versatile mathematical tool. The unifying basis of analytic geometry, and of the tremendous segment of mathematics which it spawned, is the idea that all points of a line correspond in a one–one fashion to all *real* numbers, which is the number system which we will ultimately develop in this course. First, however, we must develop some geometric concepts and appropriate language and symbolism.

## GEOMETRY AND SETS

The invention of numbers, arithmetic, and algebra made it possible for man to analyze and discuss the *quantitative* aspect of his physical universe. We might say that geometry helps him to analyze and describe the shapes and positions of objects in that universe. Just as numbers in arithmetic are abstract concepts which describe a certain property of physical objects, the elements of geometry are abstractions which describe other properties of physical objects. And, just as elements and sets were represented by both picture and letter symbols, so are the elements and sets of geometry.

## POINTS AND SETS

The basic elements of geometry are *points*. Theoretically, a point cannot be seen, because it is smaller than any physical object. In practice, a dot is used as the picture of a point, and a letter of the alphabet is written beside it to identify the point in verbal or written discussion. Both a set of points and its picture are referred to as a geometrical *figure*. It is customary to represent points by capital letters and sets of points by small letters. However, since this custom has been superseded by the more general theory of sets, we shall use small letters for points and capital letters for sets of points, in agreement with the usual set notation.

The universal set of geometry, the set of *all* points, is called *space*. The concept of space is represented by the physical space in which we live—our universe. Physical space includes not only the space in which astronauts and satellites travel, but our atmosphere, the earth—and even our own bodies. We can, in a sense, see a set of points in a book lying on the table, but the vacant place left when the book is moved is still the same set of points, occupied by the air which replaced the book. In other words, points exist everywhere, whether or not there is a physical object there. Points are *not* physical objects which can be moved around.

## POINTS AND DISTANCE

The most basic geometric property, recognized very early in life, is *distance*. To show how intuitive and elementary this idea is, the following questions are posed as exercises to be solved *before* doing any further reading. Use only a drawing compass to assist you in the exercises.

*Exercise 5.3*

1. From the picture, list the points which are (a) the same distance from *o* that *a* is; (b) closer to *o* than *a* is; (c) farther from *o* than *a* is.

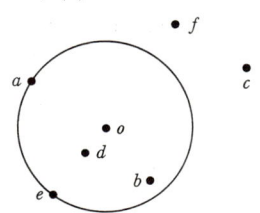

2. Show that point *b* at right is farther from point *c* than from point *d*.

3. Points *b*, *c*, and *d* are points of a circle as pictured. Which of points {*c*, *d*} is farther from point *b*? Try to locate a point of the circle which is farther from *b* than any other point of the circle.

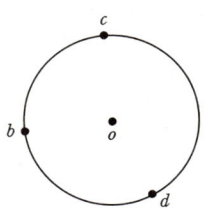

4. Find a point of the circle which is the same distance from *b* as *c* is. Try to locate a point of the circle which is closest to point *b*.

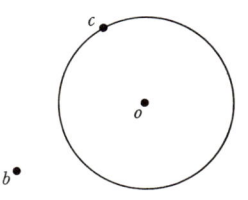

5. Find a point which is the same distance from both *b* and *c*. Repeat the procedure to find a second and third such points.

# Elementary Mathematical Structure

From these exercises it should be clear that we have an intuitive notion about equality and inequality of distances between points. The drawing compass helps to illustrate these relationships, but does not furnish us with a definition. However, since we have some indication of a common understanding of *distance*, we shall assume the word as being *undefined but understood*.

## Axiom G1

For each pair of points $b$, $c$ there is a unique *distance* $|bc|$.

Since $c$ and $b$ are the same points as $b$ and $c$, the symbol $|cb|$ must represent the same distance or, by the definition of the equals symbol,

$$|bc| = |cb|$$

## POINTS AND BETWEENNESS

Problems 3 and 4 of Exercise 5.3 illustrate a very special and useful relation between points. In finding the points of the circles farthest and closest to $b$, respectively, you may have wished for a *straight edge* (or ruler). By aligning the edge of this instrument with points $o$ and $b$, it would have been easy to show that the required points also aligned with the edge of the ruler. This will be the allowed use of the ruler in our geometry; i.e., to align it with not more than two dots, and locate others along its edge. When three points are aligned, exactly one of them

•           •           •
$a$         $b$         $c$

shall be recognized as being *between* the other two. In this illustration, the point $b$ is between points $a$ and $c$, written

$$a(b)c \quad \text{or} \quad c(b)a$$

The relation *between* shall be our third undefined term, the others being *point* and *distance*.

## Axiom G2

If $a(b)c$ then neither $b(a)c$ nor $b(c)a$; i.e., of three distinct points, only one can be between the others.

Before reading further, do the following exercise to check your understanding of the ideas discussed thus far. Compass and straight edge may be used.

*Addition of Whole Numbers*

*Exercise 5.4*

1. From the figure, write all the *between* relations shown.

       a    b  c    d

2. Find all points which belong to the set $\{p : |ap| = |bc|\}$.

     b   c
   
               a
   
   b
   
   a

3. Find all points of the set
   $\{p : |pa| = |cd| \text{ and } |pb| = |ef|\}$.

   c    d    e    f

4. Find all points of the set $\{p : a(p)b\}$.

   a        b

5. Find all points of the set $\{p : a(b)p\}$.

   a    b

Problems 4 and 5 introduce two very important types of sets, which we will define formally.

*Definition 5.8.* The set $\overline{ab} = \{a, b\} \cup \{p : a(p)b\}$ is called a line *segment*. It consists of two distinct points and all points between them. Points $a$ and $b$ are called *end points*.

*Definition 5.9.* The set $\overrightarrow{ab} = \overline{ab} \cup \{p : a(b)p\}$ is called a *ray*. Point $a$ is called the *end* point of the ray $\overrightarrow{ab}$.

       a               b            c             d

        Segment $ab$          Ray $cd$

## DISTANCES IN A RAY

Geometry, like all mathematics, is a system. To build the system would take more time than we have, but some attention should be paid to its logical structure. Therefore, on our way to the development of the number line, we shall continue to outline such a structure with a few key axioms and definitions, together with explanatory remarks. We have already shown the existence of an intuitive concept of inequality of distances similar to that for numbers; now we can make a formal definition.

84       *Elementary Mathematical Structure*

*Definition 5.10.*     For $c \neq d$, $|ab|$ is *greater than* $|cd|$ (written $|ab| > |cd|$) when there are points $p$ and $q$ for which $a(p)b$ with $|ap| = |cd|$ and $c(d)q$ with $|cq| = |ab|$. For $c = d$, $|ab| > |cd|$ when $a \neq b$ and $|ab| = |cd|$ when $a = b$.

From Definition 5.10 it follows immediately that when $a(b)c$, $|ac| > |ab|$ and $|ac| > |bc|$. The situation $|ab| > |cd|$ where $c \neq d$ is pictured below.

```
a         p   b            c         d    q
•         •   •            •         •    •
```

One of the fine points of logic, debated by mathematicians for years, is the question of whether (1) mathematical objects are already in existence, to be recognized simply by a definition; or (2) they are inventions whose existence must be formally assumed by an axiom. We are inclined toward the second viewpoint; hence the axiom about the existence of *distance* and the following axiom which accounts for the existence of more points on a ray.

## Axiom G3

For every $a \neq b$ and $c \neq d$ there is a unique point $p$ for which $a(b)p$ and $|bp| = |cd|$.

The picture below illustrates this axiom. An especially useful set of points is

```
      c    d        a          b    p
      •    •        •          •    •
```

pictured below which uses $|ab|$ as the distance $|cd|$ to locate point $e$ (for $p$); then $b$ and $e$ play the roles of $a$ and $b$ to locate point $f$; etc. This procedure

```
     a    b    e    f    g
     •————•————•————•————•————•————•————•———→
```

gives us a set of equally spaced points with an order like that of the elements of $W$. Indeed, there is a first point ($a$) and each point has a unique successor; thus we can show a one–one correspondence as pictured on part of a

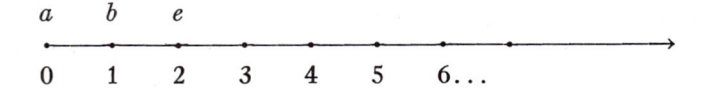

```
     a    b    e
     •————•————•————•————•————•————•————•———→
     0    1    2    3    4    5    6...
```

*number line.* Although a *ray* is pictured, the only points of immediate interest are those corresponding to the numbers under study. The number line is used in the elementary grades to demonstrate certain properties of numbers and operations with numbers.

## Addition of Whole Numbers

**Definition 5.11.** $|ef| = |ab| + |cd|$ ($a \neq b$ and $c \neq d$) iff there is a point $p$ for which $e(p)f$, $|ep| = |ab|$ and $|pf| = |cd|$. $|ef|$ is called the *sum* of $|ab|$ and $|cd|$.

It should be noted that no axiom is needed here, for if there are two points $b \neq c$ then there is a unique point $p$ (by Axiom G3) for which $b(c)p$ and $|cp| = |bc|$, so that $|bc| + |cp| = |bp|$ and a sum exists.

$$\underset{b \quad\quad c \quad\quad p}{\bullet \quad\quad \bullet \quad\quad \bullet \longrightarrow}$$

Except for one modification, we are ready to demonstrate the existence on any ray $\overrightarrow{ab}$ of a unique point $p$ such that $|ap|$ is the sum of any two specified distances. Axiom G3 provides for finding $q \in \overrightarrow{ab}$ so that $|bq|$ is the required distance, but how to start at point $a$? First, start with $\overrightarrow{ba}$ rather than $\overrightarrow{ab}$, and by Axiom G3, locate a point $h$ for which $b(a)h$ and $|ah| = |ab|$. Then use ray

$$\underset{h \quad a \quad b \quad q \quad p}{\bullet \quad \bullet \quad \bullet \quad \bullet \quad \bullet \longrightarrow}$$

$\overrightarrow{ha}$ to locate $q$ with $h(a)q$ and $|aq|$ as one of the given distances. Then use $\overrightarrow{aq}$ to locate $p$ with $a(q)p$ and $|qp|$ the other distance. Then $|ap| = |aq| + |qp|$ by Definition 5.11. This argument illustrates how even such simple ruler-and-compass constructions can represent purely geometrical ideas.

## NUMBER LINE ADDITION

The previous discussion has given us a method for adding on the number line. We will illustrate with the sum $3 + 4$. Given a number line (ray) $\overrightarrow{ab}$ as shown,

$$\underset{0 \quad 1 \quad 2 \quad 3 \quad 4 \quad 5 \quad 6 \quad 7 \quad 8}{\overset{a \quad b \quad\quad c \quad\quad\quad\quad d}{\bullet \quad \bullet \quad \bullet \quad \bullet \quad \bullet \quad \bullet \quad \bullet \quad \bullet \quad \bullet \longrightarrow}}$$

with a point $d$ such that $a(c)d$ and $|cd| = |ab| + |ab| + |ab| + |ab| = 4|ab|$, it follows that

$$|ad| = |ac| + |cd|$$
$$= 3|ab| + 4|ab|$$
$$= (3 + 4)|ab|$$
$$= 7|ab|$$

where the 3, 4 and 7 come from counting segments as elements of sets. However, the number line has an advantage in that the segments are already numbered from zero. Thus, if one end of the "sum-segment" is at 0, the other end has the required sum as its number, or *coordinate*. Note that 7 is the coordinate of point $d$.

If we identify the seven segments of the preceding example by the coordinates of their end points ($[0, 1] = \overline{ab}$, for example), we have $[0, 1]$, $[1, 2]$, $[2, 3]$, $\cdots$, $[5, 6]$, and $[6, 7]$. It appears that the coordinates of the *right* end points serve to *count* the segments just as the sets $\{1, 2, 3\}$ and $\{4, 5, 6, 7\}$ of cardinality 3 and 4 respectively are especially helpful in finding the sum $3 + 4$. In other words, although the standard sets $N_3 = \{1, 2, 3\}$ and $N_4 = \{1, 2, 3, 4\}$ are helpful in illustrating the problem of adding 3 and 4, they are not very helpful in solving that problem. In a similar way, the picture below

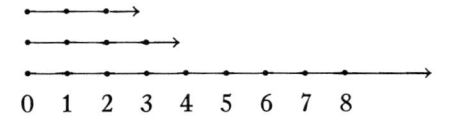

states the addition problem, $3 + 4$, but the following picture states it and shows the solution as well. (The arrows at the ends of the segments do not indicate rays, but serve to guide the viewer's eye to the terminal point.)

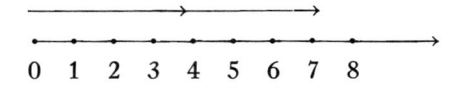

## SUBTRACTION ON THE NUMBER LINE

By considering the units in a line segment as elements of a set, it is also possible to perform subtraction. To find the result of $6 - 4$ using sets, we might use $N_6 = \{1, 2, 3, 4, 5, 6\}$ and $N_4 = \{1, 2, 3, 4\}$ since $N_4 \subseteq N_6$, and *count* the elements of $N_6 - N_4$. We would be smarter however, to use $N_6 - \{3, 4, 5, 6\} = \{1, 2\} = N_2$ since $\{3, 4, 5, 6\} \sim N_4$, and the elements of $N_6$ which remain after subtracting give us the cardinal number without our having to count.

Thus, in subtracting a segment of 4 units from a segment of 6 units, we shall be careful to leave the remaining segment at the end of the ray. In Figure 5.6, $\overline{ac}$ is 6 units long, while $\overline{cp} \subset \overline{ac}$ is 4 units long. Then $\overline{ac} - \overline{cp} =$

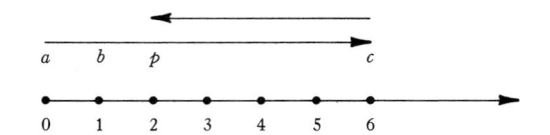

**Figure 5.6.** $6 - 4$ *on the Number Line.*

*Addition of Whole Numbers* 87

$\overline{ap}$, whose length is given by the coordinate 2 of point $p$. This procedure amounts to (1) measuring 6 units from $a$, (2) returning 4 units toward $a$, and (3) reading the coordinate of the final point. Remember: this is not a skill to be developed nor a rule to memorize. It is a device to reinforce comprehension of addition and subtraction, and is especially useful in giving meaning to negative numbers.

*Exercise 5.5*

1. Draw a picture of a ray $\overrightarrow{ab}$. Use your drawing compass to locate points $p$ and $q$ in $ab$ for which $|ap| = |cd| + |de|$ and $|aq| = 4|ce|$.

2. From the figure, list (a) all segments and (b) all rays which are shown, and which can be represented by the given letters.

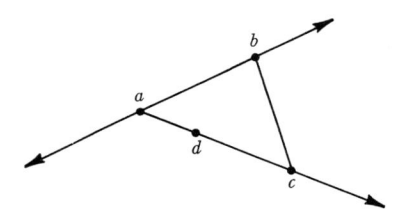

3. In the figure, $a(b)c$ is the only between relation for the points $a$, $b$, $c$ and $d$. Identify the figure represented by each of the following sets, even though it may have not been mentioned in the text.
   (a) $\overrightarrow{ab} \cap \overrightarrow{ba}$
   (b) $\overrightarrow{ba} \cup \overrightarrow{bc}$
   (c) $\overline{ac} \cup \overline{cd} \cdot \overline{ad}$
   (d) $\overrightarrow{bc} \cup \overrightarrow{bd}$
   (e) $\overrightarrow{ab} \cup \overrightarrow{bc}$

4. Draw a number line (ray $\overrightarrow{ab}$), marking consecutive points a distance $|cd|$ apart. Give the coordinate of the point $p$ for which
   (a) $|ap| = 5|cd|$
   (b) $|ap = 8|cd|$
   (c) $|ap| = 0 \cdot |cd|$

5. Illustrate the following sums on number lines:
   (a) $2 + 2 + 2$
   (b) $5 + 2$
   (c) $2 + 5$
   (d) $37 + 42$
   (e) $5 + 0$

6. From observations similar to parts (d) and (e) of Problem 5, discuss some of the inadequacies of the number line.

7. Illustrate the sum 5 + 2 using
   (a) any two sets of the proper cardinalities
   (b) two subsets of $N$ whose union demonstrates the sum as graphically as does Problem 5(b).

8. The segment [0, 1] on the number line is often called the *unit* segment and is said to be of *length* 1. [0, 2] is of length 2. What are the lengths of the following segments?
   (a) [0, 5]  (b) [0, 8]  (c) [2, 7]  (d) [4, 5]  (e) [m, n]

9. Illustrate the following operations on the number line:
   (a) 8 − 3  (b) 4 − 3  (c) 5 − 5  (d) (7 − 3) + 3  (e) (7 + 3) − 3

10. Illustrate the difference 6 − 2 by using
    (a) any set and subset of the proper cardinalities
    (b) a standard counting set and a subset whose difference demonstrates the problem most graphically.

# 6

# Multiplication of Whole Numbers

Traditionally, multiplication has been introduced in the elementary grades as a *short cut to repeated addition*. Although this is a natural and useful concept, it has some pedagogical limitations, so we will postpone investigation of this viewpoint and use another for our basic definition. The viewpoint we have selected has numerous applications in more advanced mathematics, especially in the area of mathematical analysis. As in the case of addition, there is an operation with sets which will enable us to define multiplication of whole numbers.

## CARTESIAN PRODUCTS OF SETS

We have already observed the usefulness of pairing elements from two sets. Perhaps you have also experimented with different ways of making such pairings. For example, with sets

$$A = \{a, b\} \quad \text{and} \quad B = \{c, d, e\}$$

the element $a \in A$ can be paired with each of the elements of $B$ to produce the set of pairs

$$\{(a, c), (a, d), (a, e)\}$$

The same procedure applied to $b \in A$ yields

$$\{(b, c), (b, d), (b, e)\}$$

These six pairings are shown in Figure 6.1, which also demonstrates the pairing of each element of $B$ with every element of $A$. The complete set of pairs listed above forms a special kind of set

$$\{(a, c), (a, d), (a, e), (b, c), (b, d), (b, e)\}$$

called the *cartesian product* of $A$ and $B$.

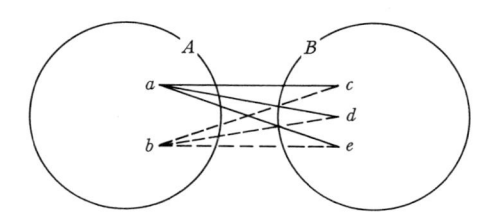

**Figure 6.1. Pairings of A × B.**

*Definition 6.1.* For any ordered pair of sets $S$ and $T$, the set $S \times T = \{(s, t) \mid s \in S, t \in T\}$ is called the *cartesian product* of $S$ and $T$.

The name of this type of set is from René Descartes, the French mathematician credited with developing the method of analytic geometry mentioned in Chapter 5. The significance of *order* should be noted in the definition, so that the set $A \times B$ of the example is different from

$$B \times A = \{(c, a), (c, b), (d, a), (d, b), (e, a), (e, b)\}$$

even though the same elements are paired. In each pair, the element belonging to the set mentioned first is the first element of the pair. Thus $(c, a) \in B \times A$ while $(a, c) \in A \times B$.

## MULTIPLYING CARDINAL NUMBERS

We have already observed that $A \times B$ (as well as $B \times A$) has six elements, while $n(A) = 2$ and $n(B) = 3$. Is it mere coincidence that

$$n(A \times B) = 6 = 2 \times 3 = n(A) \times n(B)$$

Of course, this question is asked in the context of your total mathematical experience outside this development. In that experience, you probably learned to think of $2 \times 3$ as "two threes," meaning "the sum of two threes" or $3 + 3$. But this should remind us of the way we built the set $A \times B$ as

$$\{(a, c), (a, d), (a, e)\} \cup \{(b, c), (b, d), (b, e)\}$$

Here we see $A \times B$ as the union of two disjoint sets, each having three elements, whose cardinal number then is $3 + 3$.

Another way to visualize $A \times B$ as having $n(A) \times n(B)$ elements is shown below. Here the elements of $A$ are listed above the columns, while elements

### A × B in table form

|   | a | b |
|---|---|---|
| c | (a, c) | (b, c) |
| d | (a, d) | (b, d) |
| e | (a, e) | (b, e) |

of $B$ are listed beside the rows, thus making it easy to segregate the elements of $A \times B$ which involve any particular element of $A$ or $B$. The mathematical significance of this form will be discussed later. The purpose of our present observations is to show you that counting the elements of $S \times T$ for any two sets $S$ and $T$ will yield the product $n(S) \times n(T)$ obtained by any mathematical means known.

*Definition 6.2.* For every ordered pair of sets $S$ and $T$, $n(S) \cdot n(T) = n(S \times T)$ and is called the *product* of $n(S)$ and $n(T)$.

Since every whole number is the cardinal number of some set, Definition 6.2 guarantees that the product $m \cdot n$ of any ordered pair of whole numbers is defined and is a whole number. $m \cdot n$ is read $m$ times $n$ and is also written in the forms $mn$, $m \times n$, $m(n)$, or $(m)(n)$. Observe also that although addition and multiplication in $W$ are related, their definitions as given here are independent. Thus the system $(W; \times)$ can be studied independently and compared with the system $(W; +)$ without the suspicion that $(W; \times)$ ought to be similar to $(W; +)$ since it was derived from it.

# THE SYSTEM $(W; \times)$

This brief look at the system $(W; \times)$ is primarily for the purpose of taking note of its properties rather than for gaining more experience in making proofs, so our discussion will generally parallel that of the previous chapter in the development of the system $(W; +)$.

Although the cartesian product of sets was defined as a binary operation, any number of sets could have been used. For example, if

$$A = \{a, b\}, \quad B = \{0, 1\} \quad C = \{y, z\}$$

then we could define

$$A \times B \times C = \{(a, b, c) \mid a \in A, b \in B, c \in C\}$$
$$= \{(a, 0, y), (a, 0, z), (a, 1, y), (a, 1, z), (b, 0, y), (b, 0, z)$$
$$(b, 1, y), (b, 1, z)\}$$

For some purposes this is an excellent generalization. As a means of defining products of cardinal numbers, however, it is unnecessary. To illustrate, the set

$$A \times (B \times C) = \{a, b\} \times \{(0, y), (0, z), (1, y), (1, z)\}$$
$$= \{[a, (0, y)], [a, (0, z)], [a, (1, y)], [a, (1, z)],$$
$$[b, (0, y)], [b, (0, z)], [b, (1, y)], [b, (1, z)]\}$$

has the same cardinality as $A \times B \times C$. In fact, it has the same ordered triples, although they are somewhat disguised by extra punctuation. Multiplication is thus defined as a binary operation by choice and not by necessity. This example also illustrates the *associativity* property in $(W; \times)$; for although $A \times (B \times C) \neq (A \times B) \times C$ (because of the different groupings) it is true that

$$A \times (B \times C) \sim (A \times B) \times C$$

so that

$$n[A \times (B \times C)] = n[(A \times B) \times C]$$

from which

$$n[A \times (B \times C)] = n(A) \cdot n(B \times C) = n(A) \cdot [n(B) \cdot n(C)]$$

$$n[(A \times B) \times C] = n(A \times B) \cdot n(C) = [n(A) \cdot n(B)] \cdot n(C)$$

and

$$n(A) \cdot [n(B) \cdot n(C)] = [n(A) \cdot n(B)] \cdot n(C)$$

The closure property and commutativity have already been mentioned, so we can now state these three properties as theorems, even though we have made only illustrations of methods of proof.

## Theorem 6.1

The set $W$ is closed under multiplication.

## Theorem 6.2

Multiplication is commutative in $(W; \times)$.

## Theorem 6.3

Multiplication is associative in $(W; \times)$.

## IDENTITY ELEMENTS

There is one more significant property which the systems $(W; +)$ and $(W; \times)$ share. In the system $(W; +)$ it was noted (Chapter 2) that the number zero has the property that $0 + k = k$ for every $k \in W$. You may recall that in outlining a proof for Theorem 2.1, sets $N_m$ and $N_n$ were replaced by sets of the same cardinality by using, instead of $N_4 = \{1, 2, 3, 4\}$, the set $\{(a, 1), (a, 2), (a, 3), (a, 4)\}$, for example; or in terms of newer symbols, if $A = \{a\}$

$$A \times N_4 = \{a\} \times \{1, 2, 3, 4\} = \{(a, 1), (a, 2), (a, 3), (a, 4)\}$$

Thus it appears that if $n(A) = 1$, then
$$n(A) \cdot n(B) = 1 \cdot n(B) = n(B)$$
which tends to verify the following theorem.

## Theorem 6.4

For every $k \in W$, $1 \cdot k = k$.

From the preceding example it should be clear that 0 and 1 play the same roles for addition and multiplication, respectively. That is, each combines, by the appropriate operation, with every element of $W$ to reproduce that element. 0 and 1 are called *identities* for addition and multiplication respectively.

*Definition 6.3.* An element $e \in S$ is called an *identity element* for the binary operation $*$ on $S$ iff $e * b = b * e = b$ for every $b \in S$.

## THE SYSTEM $(W; +, \times)$

A logical development of an algorithm for multiplication involves both addition and multiplication. Therefore, let us collect the properties of systems $(W; +)$ and $(W; \times)$ into the more complex system $(W; +, \times)$. In addition to retaining all the separate properties of $+$ and $\times$, we will gain the bonus of a very useful property involving both operations.

First, we will illustrate the principle with the analogous set operations $\times$ and $\cup$. In Table 13 we see $\{a, b, c, d, e\} \times \{y, z\}$

**Table 13. $\times$ Distributive Over $\cup$.**

|   | a | b | c | d | e |
|---|---|---|---|---|---|
| y | (a, y) | (b, y) | (c, y) | (d, y) | (e, y) |
| z | (a, z) | (b, z) | (c, z) | (d, z) | (e, z) |

where $\{a, b, c, d, e\} = \{a, b\} \cup \{c, d, e\}$. The recognition of $\{a, b, c, d, e\}$ as the union of two disjoint sets also reveals the cartesian product as the union of two disjoint sets
$$\{(a, y), (b, y), (a, z), (b, z)\}$$
and
$$\{(c, y), (d, y), (e, y), (c, z), (d, z), (e, z)\}$$
which are also recognizable as
$$\{a, b\} \times \{y, z\} \quad \text{and} \quad \{c, d, e\} \times \{y, z\}$$
In other words, Table 13 demonstrates the relation
$$\{a, b, c, d, e\} \times \{y, z\} = (\{a, b\} \times \{y, z\}) \cup (\{c, d, e\} \times \{y, z\})$$

In terms of cardinal numbers, Table 13 shows there are $5 \times 2$ elements in $\{a, b, c, d, e\} \times \{y, z\}$, $2 \times 2$ elements in $\{a, b\} \times \{y, z\}$, and $3 \times 2$ elements in $\{c, d, e\} \times \{y, z\}$, as well as the fact that

$$5 \times 2 = (2 \times 2) + (3 \times 2)$$

or, since $5 = 2 + 3$,

$$(2 + 3) \times 2 = (2 \times 2) + (3 \times 2)$$

This analysis indicates that the multiplication of the sum $3 + 2$ by 2 can be *distributed* to be performed by multiplying 3 and 2 individually by 2 and then adding the results. This multiplication is usually shown in reverse order; that is,

$$2 \times (3 + 2) = (2 \times 3) + (2 \times 2)$$

which is certainly equivalent to the other statement since $\times$ and $+$ are commutative in $(W; +, \times)$.

*Definition 6.4.* In a system $(S; *, @)$ the operation $*$ is *distributive over* the operation $@$ when for all $a, b, c \in S, a * (b @ c) = (a * b) @ (a * c)$.

From the analysis of our example we would suspect that the set operation $\times$ is distributive over set union. The following theorem states the case for our system $(W; +, \times)$.

## Theorem 6.5

Multiplication is distributive over addition in $(W; +, \times)$.

As you might suspect, applications of this property occur quite frequently. Perhaps for this reason, and to simplify notation, it is customary to eliminate the parentheses in the "distributed" form of

$$a \times (b + c) = (a \times b) + (a \times c)$$

to give

$$a \times (b + c) = a \times b + a \times c.$$

This means that we must agree to the convention that *when both multiplication and addition are shown without grouping symbols, the multiplications are to be performed before the additions.*

EXAMPLE: $2 + 3 \times 4 + 2 \times 3 \times 5 = 2 + 12 + 30 = 44$. However, $(2 + 3) \times 4 + 2 \times 3 = 5 \times 4 + 2 \times 3 = 20 + 6 = 26$, since the parentheses indicate that 4 is to be multiplied by the sum $2 + 3$.

*Exercise 6.1*

1. Show $A \times B$ and $B \times A$ if (a) $A = \{a, b\}$, $B = \{b, c\}$; (b) $A = \{0, 1, 2,\} = B$.
2. Find $n(A) \cdot n(B)$ and $n(B) \cdot n(A)$ for each part of Problem 1.

3. $4 \cdot 3 = 3 + 3 + 3 + 3$ can be described as "the sum of four three's." How would you similarly describe (a) $1 \cdot 3$? (b) $0 \cdot 3$?

4. Use appropriate sets and Definition 6.2 to find the cardinal number represented by (a) $4 \cdot 3$; (b) $1 \cdot 3$; (c) $0 \cdot 3$.

5. Make a list of the properties of $(W; +, \times)$ and give an example of each.

6. (a) Make a table of $\{0, 1, 2, 3\} \times \{0, 1, 2\}$ similar to that in Table 13. (b) Do the same for $\{0, 1, 2\} \times \{0, 1, 2, 3\}$.

7. (a) If $A = \{a, b\}$, $B = \{b, c\}$ and $C = \{c, d\}$, is $A \cap (B \cup C) = (A \cap B) \cup (A \cap C)$? In general, do you believe that set intersection is distributive over set union? (b) For the sets of part (a), examine the statement $A \cup (B \cap C) = (A \cup B) \cap (A \cup C)$. Describe this property in words.

8. Calculate the following to recognize the whole numbers represented, and compare problems and results.
(a) $(6 \cdot 3) + (6 \cdot 7)$     (d) $(2 + 3) \cdot (2 + 4)$
(b) $6 \cdot (3 + 7)$     (e) $(2 + 3) 2 + 4$
(c) $2 + 3 \cdot 2 +$

9. Name the property of $(W; +, \times)$ which justifies each of the following statements, assuming that $a, b, c \in W$.
(a) $(7 \times 4) \times 25 = 7 \times 100$
(b) $3 \cdot a + 4 \cdot a = 7 \cdot a$
(c) $ac + b(a + c) = b(a + c) + ac$
(d) $b(a + c) + ac = (ba + bc) + ac$
(e) $(ba + bc) + ac = ba + (bc + ac)$
(f) $ba + (bc + ac) = ab + (cb + ca)$
(g) $ab + (cb + ca) = ab + c(b + a)$
(h) $ab + c(b + a) = ab + c(a + b)$

*10. (a) Use Definition 6.3 to show that $k \cdot 0 = 0$ for every $k \in W$.
(b) Show that the identity element of a system $(S; *)$ is unique. Hint: If $a$ and $b$ are both identity elements for $*$, what does $a * b$ represent?

## PLACE VALUE IN MULTIPLICATION

By making use of the basic principles of addition in $(W; +)$ and the complementary properties of our place value system of numeration, we were able to devise an algorithm for addition which required a relatively small set of memorized *addition facts*—the sums of pairs of one-digit numbers. A similar set of basic multiplication facts are sufficient for effecting multiplication for all pairs of whole numbers, but the algorithm is a little more complicated. In fact, although a number of devices for performing multiplication were available in the Middle Ages, it was customary for merchants to consult specially prepared books of multiplication tables in computing their customer's bills. Here it will be assumed that you know the basic multiplication facts from your grade school experience, refreshed by occasional later usage. Even such "overlearned" facts will eventually fade from memory, however, so you should test yourself and relearn any facts you may have forgotten.

As in the addition process, the basic approach to a digit–by–digit multiplication procedure is through the expanded form numeration. We shall demonstrate the method by performing the multiplication 23 × 415. In order to simplify the illustration, a *semi-expanded* form consisting of the sum of the numbers represented by each digit will be used initially. In this form,

$$415 = 400 + 10 + 5$$

so that

$$23 \times 415 = 23 \times (400 + 10 + 5)$$
$$= 23 \times 400 + 23 \times 10 + 23 \times 5$$

where an extended form of distributivity of multiplication over addition has been used. We continue by expanding 23 and employing distributive, associative, and commutative properties.

$$23 \times 415 = (20 + 3) \times 400 + (20 + 3) \times 10 + (20 + 3) \times 5$$
$$= 20 \times 400 + 3 \times 400 + 20 \times 10 + 3 \times 10 + 20 \times 5$$
$$+ 3 \times 5$$
$$= (2 \times 10) \times (4 \times 100) + 3 \times (4 \times 100) + (2 \times 10)$$
$$\times (1 \times 10) + 3 \times (1 \times 10) + (2 \times 10) \times 5 + 3 \times 5$$
$$= (2 \times 4)(10 \times 100) + (3 \times 4)(100) + (2 \times 1)(10 \times 10)$$
$$+ (3 \times 1)(10) + (2 \times 5)(10) + 3 \times 5$$
$$= (2 \times 4)(10 \times 100) + (2 \times 1)(10 \times 10) + (2 \times 5)(10 \times 1)$$
$$+ (3 \times 4)(1 \times 100) + (3 \times 1)(1 \times 10) + (3 \times 5)(1 \times 1).$$

In analyzing this product, you should observe that the number represented by each digit of 23 is multiplied by each number represented by the digits of 415 and these products are all added together. In the last form each digit of 23 is multiplied by each digit of 415 and these products are each multiplied by the place values of the two digits. For example, in the product

$$(2 \times 4)(10 \times 100)$$

2 and 4 are the digits while 10 and 100 are their respective place values. In order to further modify and simplify the form of our result, we pause to comment on the nature of products of place values.

To say that $10 \times 10 = 100$ is well known would ignore our agreement to limit our knowledge of multiplication facts to products of one-digit numbers. The fact could be verified by using Definition 6.2 to calculate $10 \times 10$, but this would be laborious and even prohibitive for $10 \times 100$ or $100 \times 1000$. The argument

$$10 \times 10 = (9 + 1) \times 10 = 9 \times 10 + 10 = 90 + 10 = 100$$

from properties of $(W; +, \times)$ and knowledge of addition, can be extended.

For example,

$$\begin{aligned}100 \times 10 &= (90 + 10) \times 10 = 90 \times 10 + 10 \times 10 \\ &= 9 \times (10 \times 10) + (10 \times 10) \\ &= 9 \times 100 + 100 \\ &= 900 + 100 \\ &= 1000\end{aligned}$$

using the prior calculation $10 \times 10 = 100$. By induction, the properties of $10^n$ can be established. We state the result we wish to use as a theorem.

## Theorem 6.6

$$10^m * 10^n = 10^{m+n} \text{ for } m, n \in W.$$

In using this theorem we must remember that $10^0 = 1$ by definition, $10^1 = 10$ and $10^n$ is the product of $n$ tens for $n \geq 2$.

Also, in decimal numeration $10^n$ is represented by the digit 1 followed by $n$ zeroes. For example,

$$10^4 = 10 \times 10 \times 10 \times 10 = 10000$$

Returning to our problem and using the preceding principles, we see that

$$\begin{aligned}23 \times 415 &= (2 \times 4)(10 \times 100) + (2 \times 1)(10 \times 10) + (2 \times 5)(10 \times 1) \\ &\quad + (3 \times 4)(1 \times 100) + (3 \times 1)(1 \times 10) + (3 \times 5) \\ &= 8 \times 1000 + 2 \times 100 + 10 \times 10 + 12 \times 100 + 3 \times 10 + 15 \\ &= 8000 + 200 + 100 + 1200 + 30 + 15\end{aligned}$$

The product is left in this form to be compared with the *partial product* form of the algorithm shown below.

$$\begin{array}{r} 415 \\ \times 23 \\ \hline 8000 \\ 200 \\ 100 \\ 1200 \\ 30 \\ 15 \\ \hline 9545 \end{array}$$

Notice that the writing of 12 × 100 as 1200 (twelve hundred) gives the correct result

$$(10 + 2) \times 100 = 10 \times 100 + 2 \times 100 = 1000 + 200$$

(one thousand two hundred) since the digit 1 is "shoved over" into its correct place automatically when the digit 2 is written in hundreds place. In the more standard form, shown below, the multiplication by the lower right digit is done first. This form is shown together with identification of the pair of digits being multiplied and the product of their place values.

|  | digits multiplied | place value |
|---|---|---|
| 415 |  |  |
| ×23 |  |  |
| 15 | 3 × 5 | one |
| 30 | 3 × 1 | ten |
| 1200 | 3 × 4 | hundred |
| 100 | 2 × 5 | ten |
| 200 | 2 × 1 | hundred |
| 8000 | 2 × 4 | thousand |
| 9545 | complete product |  |

Another rather interesting algorithm for multiplication which makes use of the partial products concept and places digits in their proper places, is the so-called *lattice* method. The product 23 ∗ 415 is demonstrated below.

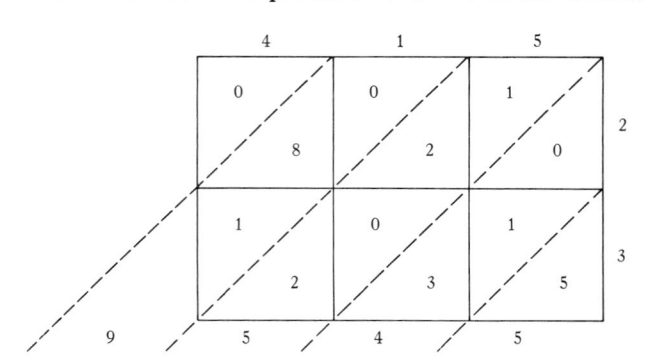

**Figure 6.2. 23 × 415 by the Lattice Method.**

Since the product of two one-digit numbers can always be expressed with two digits (for example, 2 × 3 = 06), each can be written in the appropriate square. By arranging the digits of 23 and 415 in the order shown, all digits of the partial products which occur in the same diagonal strip have the same place value. Furthermore, the place values are arranged in proper sequence so that the sums of the digits in each strip give the final product as shown on the lower row of digits.

*Multiplication of Whole Numbers* 99

## NON-DECIMAL BASES

Since the place values of a numeral in any base are 1, $10_{base}$, $10^2_{base}$, etc., when expressed in that base, the foregoing theory regarding *place value* will be the same in every base. That is, the algorithm regarding digit–by–digit multiplication and the placement of digits in partial products is still valid. The only difference will be in the digits used to express the partial products. This calls for multiplication and addition tables. You will be given tables in bases four and seven, with an example in each base.

EXAMPLE in Base Four: $23_{four} \times 312_{four}$

| + | 0 | 1 | 2 | 3 |
|---|---|---|---|---|
| 0 | 0 | 1 | 2 | 3 |
| 1 | 1 | 2 | 3 | 10 |
| 2 | 2 | 3 | 10 | 11 |
| 3 | 3 | 10 | 11 | 12 |

| × | 0 | 1 | 2 | 3 |
|---|---|---|---|---|
| 0 | 0 | 0 | 0 | 0 |
| 1 | 0 | 1 | 2 | 3 |
| 2 | 0 | 2 | 10 | 12 |
| 3 | 0 | 3 | 12 | 21 |

```
   312
 ×23 four
```

|  | digits multiplied | place value |
|---|---|---|
| 12 | 3 × 2 | one |
| 30 | 3 × 1 | four |
| 2100 | 3 × 3 | sixteen |
| 100 | 2 × 2 | four |
| 200 | 2 × 1 | sixteen |
| 12000 | 2 × 3 | sixty-four |

$21102_{four}$

One way to check such work is to convert the original numerals to base ten, multiply, and then convert that result to the other base. Thus $312_{four} = 3 \times 16 + 1 \times 4 + 2 = 54$ and $23_{four} = 2 \times 4 + 3 = 11$, whence $54 \times 11 = 594$. Then

```
4| 594
 4| 148—2
  4| 37—0
   4| 9—1
    4|2—1
     0—2,  which checks
```

EXAMPLE in Base Seven: $36_{seven} \times 42_{seven}$

| + | 0 | 1 | 2 | 3 | 4 | 5 | 6 |
|---|---|---|---|---|---|---|---|
| 0 | 0 | 1 | 2 | 3 | 4 | 5 | 6 |
| 1 | 1 | 2 | 3 | 4 | 5 | 6 | 10 |
| 2 | 2 | 3 | 4 | 5 | 6 | 10 | 11 |
| 3 | 3 | 4 | 5 | 6 | 10 | 11 | 12 |
| 4 | 4 | 5 | 6 | 10 | 11 | 12 | 13 |
| 5 | 5 | 6 | 10 | 11 | 12 | 13 | 14 |
| 6 | 6 | 10 | 11 | 12 | 13 | 14 | 15 |

| × | 0 | 1 | 2 | 3 | 4 | 5 | 6 |
|---|---|---|---|---|---|---|---|
| 0 | 0 | 0 | 0 | 0 | 0 | 0 | 0 |
| 1 | 0 | 1 | 2 | 3 | 4 | 5 | 6 |
| 2 | 0 | 2 | 4 | 6 | 11 | 13 | 15 |
| 3 | 0 | 3 | 6 | 12 | 15 | 21 | 24 |
| 4 | 0 | 4 | 11 | 15 | 22 | 26 | 33 |
| 5 | 0 | 5 | 13 | 21 | 26 | 34 | 42 |
| 6 | 0 | 6 | 15 | 24 | 33 | 42 | 51 |

$$\begin{array}{r} 42 \\ \times 36_{seven} \\ \hline 15 \\ 330 \\ 60 \\ 1500 \\ \hline 2235_{seven} \end{array} \quad \begin{array}{l} (6 \times 2) \times 1 \\ (6 \times 4) \times 10_{seven} \\ (3 \times 2) \times 10_{seven} \\ (3 \times 4) \times 10^2_{seven} \end{array}$$

*Exercise 6.2*

1. Calculate in expanded form:
   (a) $32 \times 23$    (b) $12_{seven} \times 32_{seven}$    (c) $43 \times 15$
2. In the product $2136 \times 405$ give the place value of the following partial products:
   (a) $2 \times 4$    (b) $3 \times 4$    (c) $1 \times 0$    (d) $3 \times 5$
3. Use the lattice method to perform the following:
   (a) $214 \times 32$    (b) $32 \times 214$    (c) $212_{four} \times 321_{four}$
4. Use the partial products method to perform the multiplications of Problem 3.
5. Calculate:
   (a) $3_{four} \times 231_{four}$    (b) $23_{four} \times 312_{four}$
6. Calculate:
   (a) $5_{seven} \times 306_{seven}$    (b) $214_{seven} \times 3012_{seven}$
7. Make an addition table for base five.
8. From the results of Problem 7, make a multiplication table for base five.
   (Remember: $2 \times 4 = 4 + 4$, $3 \times 3 = 3 + 3 + 3$, etc.)

9. Use your tables from Probem 8 to perform the following in base five:
   (a) $20_{five} \times 312_{five}$   (b) $42_{five} \times 232_{five}$

## MULTIPLICATION AND GEOMETRY

In Chapter 5 we observed that one line segment can sometimes be measured by another in terms of a natural number. From this simple concept the number line, or ray, was developed as an aid for demonstrating addition and subtraction of whole numbers. Since multiplication by a natural number can be interpreted as repeated addition, it can also be demonstrated on the number line. The figure below illustrates the product

$$3 \times 2 = 2 + 2 + 2$$

on the number line, where $|ac| = 2|ab|$

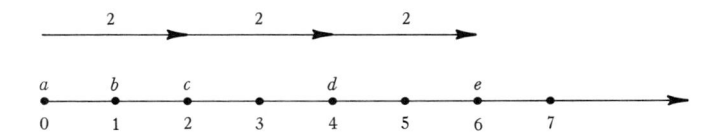

and
$$|ae| = |ac| + |cd| + |de|$$
$$= 2|ab| + 2|ab| + 2|ab|$$
$$= (2 + 2 + 2)|ab|$$
$$= (3 \times 2)|ab|$$

As in the case of addition, the coordinate of point $e$ identifies the product 6. The arrows above the number line help to illustrate visually the sum

$$2 + 2 + 2$$

In the table on p. 90, we demonstrated an arrangement of the elements of $A \times B$ which was also somewhat geometric in nature. In order to identify these ordered pairs as points, we will use two number lines arranged as pictured below. Two points on the horizontal ray are identified with the elements of

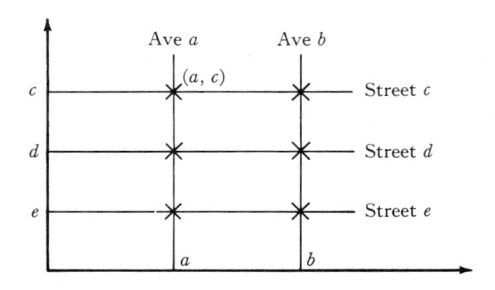

$\{a, b\} = A$, while three points on the vertical ray are labeled $c, d, e$ according to the elements of $\{c, d, e\} = B$. Thinking of the two lighter vertical lines as avenues on a city map where the three lighter horizontal lines represent streets, we would describe the $2 \times 3 = 6$ points marked as intersections of avenues and streets. Thus the point labeled $(a, c)$ would be described as "the intersection of Avenue $a$ with Street $c$."

Figure 6.3 shows a portion of the *graph* of the set
$$W \times W = \{(0, 0), (0, 1), \cdots, (1, 0), \cdots, (2, 3), \cdots\}$$
as it is used in *analytic geometry*. Notice that points on the number lines them-

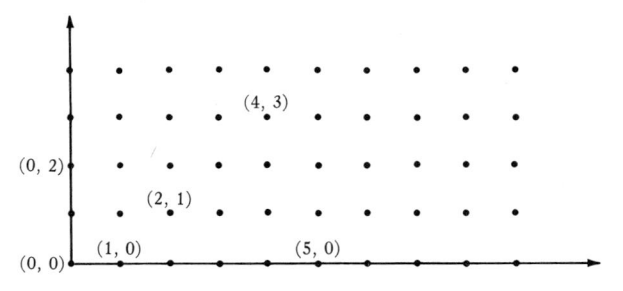

**Figure 6.3.** *Graph of* $W \times W$.

selves are now identified by ordered pairs of elements of $W$ since the two rays share the common end point which has zero as a coordinate in each ray. In order to avoid confusion, only a few of the points are labeled.

## PLANE GEOMETRY

The woman who buys material to make a dress and the painter who buys paint to paint a house have a common problem. The amount of material or paint depends upon the amount of *surface* which they must cover.

Before discussing methods of measuring portions of a surface, we will examine the nature of the thing we propose to measure. In the process, we will identify several familiar figures which make it easier to describe some of the properties with which we shall be concerned.

*Definition 6.5.*   A *line* $\overleftrightarrow{ab}$ is the set $\overrightarrow{ab} \cup \overrightarrow{ba}$.

This definition implies that a line consists of two distinct points together with all points between them and all points beyond them in each direction. The union of these sets is described below in picture and in set symbols.

$$\overleftrightarrow{ab} = \overline{ab} \cup \{p \mid p(a)b\} \cup \{q \mid a(b)q\}$$

Suppose $a$, $b$, $c$ represent three distinct points such that $c \notin \overleftrightarrow{ab}$ as shown in Figure 6.4. Then the figure $\overrightarrow{ba} \cup \overrightarrow{bc}$ is called an *angle*.

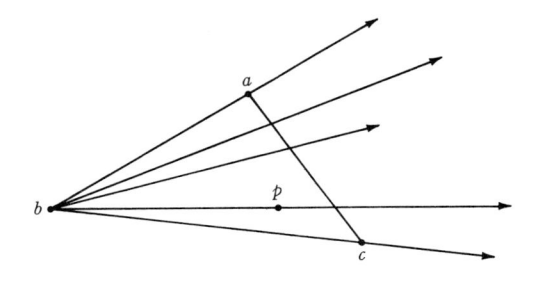

**Figure 6.4.   Angle $\widehat{abc}$ and its Interior.**

*Definition 6.6.*   An *angle* $\widehat{abc} = \overrightarrow{ba} \cup \overrightarrow{bc}$. $b$ is its *vertex*; $\overrightarrow{ba}$ and $\overrightarrow{bc}$ are its *sides*. $\widehat{aba}$ is called a *zero* angle; $\widehat{abc}$ is a *straight* angle iff $a(b)c$. $\widehat{abc}$ is a *proper* angle iff it is neither zero nor straight.

*Definition 6.7.*   A point is *interior* to a proper angle $\widehat{abc}$ iff $\overrightarrow{bp}$ intersects $\overline{ac}$ in a point between $a$ and $c$. The *interior* of a proper angle is the set of all points interior to the angle.

Because of the property described by Axiom G3 of Chapter 5, a proper angle $\widehat{abc}$ implies the existence of many points other than those mentioned in the preceding definitions. Figure 6.5 illustrates how the distinct lines $\overleftrightarrow{ab}$ and $\overleftrightarrow{bc}$

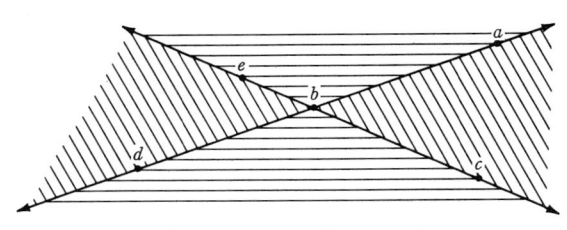

**Figure 6.5.   Plane abc.**

define four proper angles and their interiors; Axiom G3 provides for points $d$ and $e$. The entire set is called a *plane surface* or simply a *plane*. In a more formal development we would make provision for the existence of all points under discussion. For our purposes, it is more appropriate to have the concept that a plane is like a boundless floor, endless and perfectly smooth. For the sake of precision, however, we can make a formal statement.

*Definition 6.8.*   A plane $abc$ consists of the four proper angles defined by distinct $\overleftrightarrow{ab}$ and $\overleftrightarrow{bc}$ together with their interiors.

Subsets of a plane are called *plane figures*, and these are the geometric figures which we can best represent with drawings since the surface of our paper is somewhat like a plane. A sampling of familiar plane figures is given in Figure 6.6. For the time being, we will be concerned with the rectangle, the square, and their interiors.

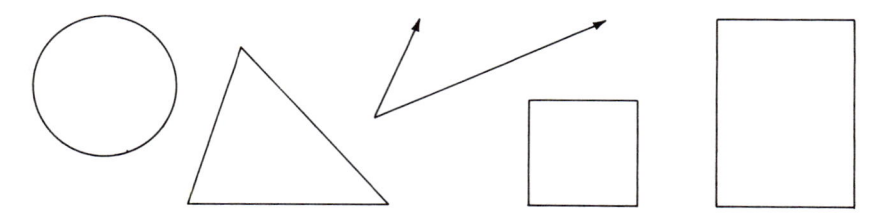

*Figure 6.6. Some Plane Figures.*

## MEASURING AREA IN SQUARE UNITS

The name *rectangle* probably comes from a German word, *recht*, meaning *right*, in the geometric sense. A *right* angle might be described as one which makes a square corner, as pictured below. Check the picture carefully against the formal definition which follows.

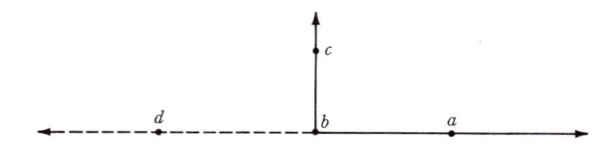

*Definition 6.9.* $\widehat{abc}$ is a *right* angle iff there is a point $d$ such that $a(b)d$, $|ab| = |bd|$ and $|ac| = |cd|$.

From the definition and illustration, it should be apparent that $\widehat{cbd}$ is also a right angle. Furthermore, if two right angles in the same plane share exactly one side, their other sides together form a straight angle. You may recall from your study of geometry that "the sum of the angles of a triangle is a straight angle (or *two* right angles)." This concept agrees with the assertion of the existence of a figure, as pictured, where the dotted segment helps present the angles of the rectangle as *sums of angles* of the two triangles. That is, the sum of the angles of a quadrilateral should be four right angles.

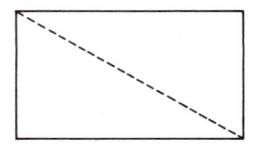

*Definition 6.10.* A *quadrilateral* $\overline{abcd}$ is a plane figure $\overline{ab} \cup \overline{bc} \cup \overline{cd} \cup \overline{da}$ in which the *sides* $\overline{ab}, \overline{bc}, \overline{cd}, \overline{da}$ share no points other than end points, called *vertices* of the quadrilateral. A *rectangle* is a quadrilateral, each angle of which is a right angle. A *square* is a rectangle whose sides are of equal length.

You should recall that pairs of opposite sides of a rectangle have the same length, as pictured in Figure 6.7 where $|ae|$ is the standard unit of length. Thus

$$|ab| = 3|ae| = |cd|$$

and

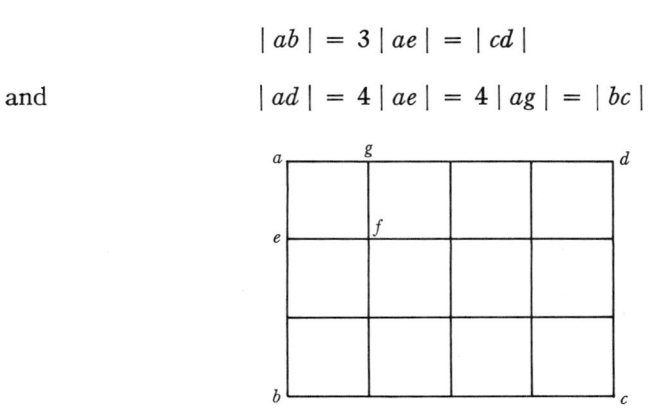

$$|ad| = 4|ae| = 4|ag| = |bc|$$

**Figure 6.7. Measuring the Area of $\overline{abcd}$.**

Since there is a one–one correspondence between end points of segments on $\overline{ad}$ and $\overline{bc}$, segments joining these pairs of corresponding points exist as shown by the additional vertical marks. Similar horizontal segments intersect the vertical segments to form

$$3 \times 4 = 12$$

squares having $|ae|$ as the lengths of their sides. If you can visualize the rectangle $\overline{abcd}$ as the outline of the floor of a small room and the solid square portions as floor tiles, the number of these tiles should seem to be a reasonable measure of the size of the room.

The portion of the plane representing the floor of the room is called a *region* of the plane. It is a subset of the plane consisting of all points inside the rectangle, and is more precisely called a *rectangular region*. The property of this rectangular region which we have measured by counting the number of *square regions* into which it was subdivided is called *area*. The area of a plane region compares to the length of a line segment as a geometric property.

From this illustration, the reason for using multiplication in the calculation of the square measure of the area of a plane region should be apparent. One reason for the use of the square region as a unit of area measurement is the fact that squares fit together nicely because of their regular shape. Other figures also have this characteristic. Some of them, which you may have seen in floor tile and masonry are shown on page 106.

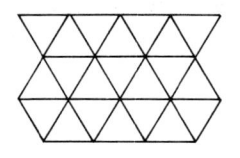

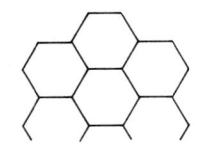

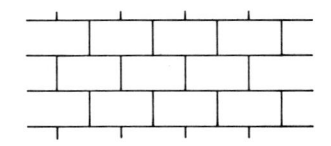

## Exercise 6.3

**1.** Illustrate the following products on number lines:
(a) $2 \times 3$  (b) $3 \times 4$  (c) $4 \times 3$

**2.** Draw a graph of (a) $A \times B$ (b) $B \times A$ where $A = \{1, 2, 3\}$ and $B = \{2, 3, 5, 6\}$.

**3.** On a graph like that in Figure 6.3, select points which represent the set $\{(x, y) \mid y = 2x\} = \{(0, 0), (1,2), (2, 4), \cdots\}$ for all $x < 7$.

**4.** Classify $\overleftrightarrow{abc}$ in each picture:

(a)  (b)  (c)

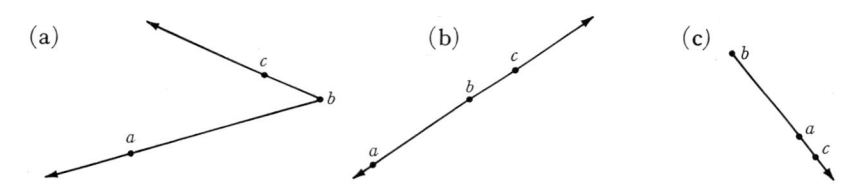

**5.** Color the set of interior points in these pictures.

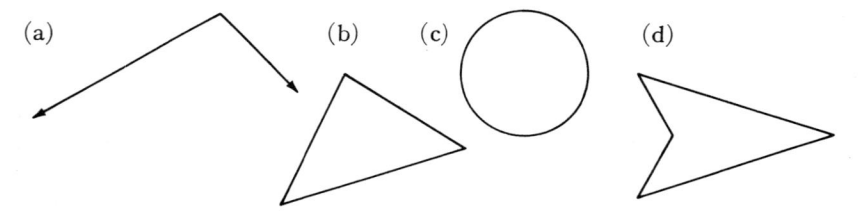

(a)  (b)  (c)  (d)

**6.** Use a ray $\overrightarrow{ba}$, a compass, and a straight edge to construct a picture of a right angle $\widehat{abe}$, using Definition 6.9 as a guide.

**7.** Which of the following represent quadrilaterals?

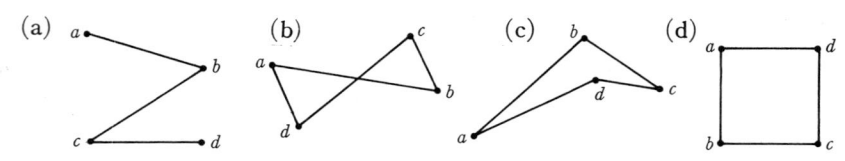

(a)  (b)  (c)  (d)

8. A rectangular room is 12 feet long and 9 feet wide. How many floor tiles will it take to cover the floor exactly if the square tiles measure (a) 12 inches on a side? (b) 9 inches on a side? (c) 6 inches on a side?

9. (a) Using the answer to part (a) of Problem 8, calculate part (c) from that answer. (b) By the same technique, predict the number of four-inch tiles which would be required to pave the floor.

10. Why do you think the square region was used to measure the area of a rectangular region rather than the triangular, hexagonal or rectangular shapes shown above?

# 7

# Inverse Operations

In working with immature minds, it is necessary to introduce abstract ideas through illustrations of concrete objects. Initially, children become familiar with numbers by counting concrete objects. To a limited extent, this principle has been applied to introduce each major mathematical concept in this book. You may have been aware, for instance, that the idea of *measure* was introduced, both for length and for area, with concrete illustrations. The exploitation of this basic principle of learning is one of the most important uses of *sets*.

In deference to this principle and to demonstrate the use of sets and set operations as a basis for all fundamental arithmetical operations, both addition and multiplication were defined by the use of sets. We have shown, however, that multiplication in $(W; +, \times)$ could have been defined in terms of addition in $(W; +)$ with no further mention of sets. The latter treatment and the Peano treatment of the system $(W; +, \times)$ are examples of a more abstract and formal definition of $(W; +, \times)$ as an *algebraic system*. Although most mathematical ideas have some intuitive or physical basis, the greatest part of mathematics has been developed from the application of more formal procedures. It is imperative that children and adults be encouraged to understand and have confidence in the *formal* aspects of mathematics. This understanding does not mean perpetuating the earlier practice of memorizing meaningless formulas and processes; it simply means *extending* the basic intuitive concepts to greater abstraction in step-by-step, logically connected, levels. The remaining material in this book has been organized with increasing emphasis on formalization, so that less time is spent in the use of set materials appropriate for the lower grades; the greater portion of the material to be discussed is for the middle and upper grades.

# SUBTRACTION

When we introduced subtraction as a *set* operation, we demonstrated the corresponding operation of subtraction of cardinal numbers. We stated at that time, however, that a different definition for subtraction would be made later. The following will be our official definition for subtraction in $(W; +, \times)$. The basis of the definition is that performed subtraction is checked by using basic addition facts.

*Definition 7.1.* If $m, n \in W$ with $m \geq n$, $m - n$ represents a number $k$ such that $k + n = m$; i.e., $m - n = k$ if and only if $k + n = m$. $m - n$ is called the *difference* of $m$ and $n$; $-$ is read *minus* and represents an algebraic operation called *subtraction*.

EXAMPLE: "$13 - 7 = 6$ because $6 + 7 = 13$" is an illustration of Definition 7.1. In the lower grades, this relation is usually introduced in the form $\_\_ + 7 = 13$ or $7 + \_\_ = 13$, where the numeral 6 is to be placed in the space. Later, it is presented as an algebraic equation, $n + 7 = 13$, from which the student is expected to conclude that $n = 6$. This close relation between addition and subtraction is sometimes described as *inverse*, and subtraction is said to be the operation *inverse* to addition.

It should be recognized that Definition 7.1 does not state that $m - n$ represents a *unique* number. Of course, if it does not, then according to our restrictions on notation we would not be allowed to use the symbol $m - n$ since a symbol cannot represent two or more distinct objects in the same context. However, if there is a number $k$ which $m - n$ represents, and if there is some other number $h$ which $m - n$ represents, then by Definition 7.1, both

$$m = h + n$$

and

$$m = k + n$$

so that by the definition of $=$ we have

$$h + n = k + n$$

Thus our question of uniqueness for subtraction is equivalent to the question, "Is it possible to add two different numbers to the same number and obtain the same sum?" We must pause to look at a special case and special forms involving subtraction, which will enable us to resolve the question.

## *Lemma*

$n - n = 0$ for every $n \in W$.

*Proof:* Since $n = n$, $n - n$ is defined by Definition 7.1 to be a number $k$ for which $k + n = n$. Since it has been shown that 0 is the only element of $W$ which adds to $n$ to give $n$, we observe that $k = 0$ or $- n = 0$.

## Lemma

$(m - n) + n = m$ for $m, n \in W$ where $m \geq n$.

(The proof is left as an exercise for the student.)

From the discussion of equality and inequality of cardinal numbers of sets in Chapter 2, it follows that

$$m + n \geq n$$

for every $m, n \in W$. Then by Definition 7.1

$$(m + n) - n = k$$

provided that

$$k + n = (m + n)$$

which is true for $k = m$. Thus, since

$$m + n = (m + n)$$

it is also true that

$$(m + n) - n = m$$

and we have another obvious but important result.

## Lemma

$(m + n) - n = m$ for all $m, n \in W$.

This last result helps us answer the question, "Does $h + n = k + n$ for $h, k, n \in W$ imply that $h = k$?" Since

$$h = (h + n) - n$$

by the lemma, then if $h + n = k + n$, substitution and another application of the lemma shows that

$$h = (h + n) - n = (k + n) - n = k$$

## Theorem 7.1

$h + n = k + n$ implies $h = k$ for all $h, k, n \in W$.

This is often called the *cancellation property* for addition. It is the basis for drawing such conclusions as found in the example following Definition 7.1. Thus, from

$$n + 7 = 13$$

we write

$$n + 7 = 6 + 7$$

and conclude that $n$ and 6 must represent the same number by Theorem 7.1; that is, $n + 7 = 13$ if and only if $n = 6$.

## SUBTRACTION ALGORITHM

As is usually the case, the formal definition does not provide a very convenient or rapid method for *performing* the subtraction operation. The basic subtraction facts are memorized just as are the addition and multiplication facts. In order to use them in algorithmic fashion with digital numerals, it will be necessary to derive a kind of *associative* and *commutative* property for subtraction.

For example, in calculating $76 - 24$, we need to know whether it is permissible to write

$$(7 \times 10 + 6) - (2 \times 10 + 4) = (7 - 2) \times 10 + (6 - 4)$$

To answer this question, we need the answer to two other questions:
Is
$$(a + b) - (c + d) = (a - c) + (b - d)$$
and is
$$(m - n)k = mk - nk$$

We shall state these principles as theorems for easy reference during the development of an algorithm for subtraction.

## Theorem 7.2

For $a, b, c, d \in W$, $(a + b) - (c + d) = (a - c) + (b - d)$ if all represented numbers are in $W$.

*Proof*: $[(a - c) + (b - d)] + (c + d) = [(a - c) + c] + [(b - d) + d]$ by the associative and commutative properties of addition in $(W; +, \times)$. But by Definition 7.1, $(a - c) + c = a$ and $(b - d) + d = b$. Thus by substitution

$$[(a - c) + (b - d)] + (c + d) = a + b$$

Also, by Definition 7.1,

$$[(a + b) - (c + d)] + (c + d) = a + b$$

or

$$[(a + b) - (c + d)] + (c + d) = [(a - c) + (b - d)] + (c + d)$$

so that

$$(a + b) - (c + d) = (a - c) + (b - d)$$

by Theorem 7.1.

## Theorem 7.3

If $m, n, k$ and $m - n$ are whole numbers, then $(m - n)k = mk - nk$.

*Proof:* $(m-n)k + nk = [(m-n)+n]k$ by the commutative and distributive properties in $(W; +, \times)$. Also, $(m-n)+n = m$ by Definition 7.1, so that

$$(m-n)k + nk = mk$$

or

$$(m-n)k = mk - nk$$

by Definition 7.1.

EXAMPLE: Now we are prepared to demonstrate the validity of the customary digit-by-digit subtraction algorithm. Using expanded form,

$$
\begin{aligned}
76 - 24 &= (7 \times 10 + 6) - (2 \times 10 + 4) \\
&= (7 \times 10 - 2 \times 10) + (6 - 4) &&\text{(Theorem 7.2)} \\
&= (7 - 2) \times 10 + (6 - 4) &&\text{(Theorem 7.3)} \\
&= 5 \times 10 + 2 \\
&= 52
\end{aligned}
$$

Of course, the expanded form was used in order to demonstrate the thought process which should occur in using the more conventional form of the algorithm,

$$
\begin{array}{r}
76 \\
-24 \\
\hline
52
\end{array}
$$

The problem is more complex if the numerals involved have more digits. Although the theorems apply immediately only to two-digit numerals, the nature of their proof should show that their results extend to numerals with any number of digits.

EXAMPLE: A second complication is illustrated by the problem $74 - 26$ where something must be done about $4 - 6$. Actually, we avoid this situation by thinking of 74 as $60 + 14$ instead of $70 + 4$. Then

$$
\begin{aligned}
74 - 26 &= (6 \times 10 + 14) - (2 \times 10 + 6) \\
&= (6 \times 10 - 2 \times 10) + (14 - 6) \\
&= (6 - 2) \times 10 + (14 - 6) \\
&= 4 \times 10 + 8 \\
&= 48
\end{aligned}
$$

In working with children, an intermediate form of partial subtraction similar

*Elementary Mathematical Structure*

to the following might be a useful device for showing the principle while avoiding some of the complications of notation inherent in the expanded form.

$$\begin{array}{r} 74 = \phantom{-}60 + 14 \\ -26 = -20 - \phantom{0}6 \\ \hline 48 + \phantom{0}8 = 48 \end{array}$$

A still more advanced algorithm uses the same principle, in a form similar to carrying in addition.

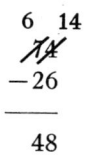

$$\begin{array}{r} \phantom{-}6\phantom{0}14 \\ \phantom{-}\not{7}\not{4} \\ -26 \\ \hline 48 \end{array}$$

## Exercise 7.1

1. Use the expanded form of numerals and the principles developed in this section to perform the following calculations.
   (a) $48 - 15$  (c) $81 - 45$
   (b) $746 - 231$  (d) $713 - 246$

2. Perform the calculations of Problem 1 using the modified or intermediate form.

3. Perform the calculations of Problem 1 using the short form of borrowing.

4. Fill in the blanks using Definition 7.1.
   (a) $11 - 3 = $ ____ because ____ $+ 3 = 11$.
   (b) $9 + 7 = 16$; therefore $16 - 7 = $ ____ and $16 - 9 = $ ____.

5. Prove that $(m - n) + n = m$ for $m, n \in W$ with $m \geq n$.

6. Find the number which $n$ represents:
   (a) $5 + n = 11$   (d) $13 - n = 9$
   (b) $n = 11 - 5$   (e) $(7 + n) - 4 = 14$
   (c) $n - 6 = 13$

*7. For each equation of Problem 5, express $n$ in terms of the other numbers using the relation between addition and subtraction. For example, since (a) $5 + n = 11$, it follows that $n = 11 - 5$.

8. Assume that $b$ and $c$ are whole numbers for which the following statements are true. In each case express $n$ in terms of $b$ and $c$.
   (a) $n + c = b$      (d) $b - (c + n) = 0$
   (b) $n - b = c$      (e) $(b - c) - n = 0$
   (c) $(n + b) - c = 0$

9. Calculate the following:
   (a) $(17 - 8) - 3$   (c) $17 - (8 + 3)$
   (b) $17 - (8 - 3)$   (d) $17 - 8 - 3$

*10. Part (d) of Problem 8 should have caused you some concern because the results of parts (a) and (b) show that subtraction is not associative. We shall agree that a horizontal sequence of numbers, separated by addition and subtraction symbols but ungrouped, shall be combined in order from left to right. For example, $34 - 19 + 6 - 15 = [(34 - 19) + 6] - 15 = [15 + 6] - 15 = 21 - 15 = 6$.

Calculate the following:
(a) $17 + 3 - 10$   (b) $23 - 17 + 3 + 5$   (c) $40 - 20 - 10 - 5$

*11. From the example of Problem 9, did you notice that the *total* results of adding 34 and 6 and subtracting 19 and 15 might be accomplished by subtracting $19 + 15$ from $34 + 6$? That is, $34 - 19 + 6 - 15 = (34 + 6) - (19 + 15) = 40 - 34 = 6$; also, $34 - 19 + 6 - 15 = 34 - (19 + 15) + 6 = 34 - 34 + 6 = 0 + 6 = 6$.

Can you prove that $a - b - c = a - (b + c)$?

## MULTIPLICATION AND DIVISION

From the intuitive viewpoint of sets, the inverse relation of addition and subtraction is well illustrated with two disjoint sets

$$A = \{a, b, c\} \quad \text{and} \quad B = \{d, e\}$$

since the facts that

$$A \cup B = \{a, b, c, d, e\}$$

and

$$(A \cup B) - B = A$$

clearly illustrate the related statements

$$n(A) + n(B) = n(A \cup B)$$

and

$$n(A \cup B) - n(B) = n(A)$$

A similar inverse relationship exists between multiplication and division, which will be illustrated by sets of symbols representing sets of traditional concrete objects.

Let $\{b, d, f\}$ represent three little boys named Bobby, Don and Frank. Suppose each boy has an apple and a cookie, where $\{a, c\}$ represent the apple and the cookie. To distinguish between the apples and cookies belonging to different boys we shall let $(b, a)$ represent Bobby's apple; $(f, c)$, Frank's cookie, etc. Then the set of apples and cookies would be represented by the set

$$\{b, d, f\} \times \{a, c\} = \{(b, a), (b, c), (d, a), (d, c), (f, a), (f, c)\}$$

Note that this set can be *partitioned* into disjoint subsets in two natural ways:

$$\{(b, a), (b, c)\} \cup \{(d, a), (d, c)\} \cup \{(f, a), (f, c)\}$$

or
$$\{(b, a), (d, a), (f, a)\} \cup \{(b, c), (d, c), (f, c)\}$$

The first partitioning is by ownership, while the second is by type of object, demonstrating the products

$$3 \times 2 \quad \text{and} \quad 2 \times 3$$

respectively. This *partitioning* of a set into equivalent disjoint subsets has been the traditional concrete introduction to the concept of division. The fact that a set of cardinality 6 can be partitioned into 3 disjoint subsets each of cardinality 2 is a way of saying that "6 divided by 3 is 2" or

$$6 \div 3 = 2$$

Similarly, the second method of partitioning illustrates the fact that "6 divided by 2 is three," or

$$6 \div 2 = 3$$

From the preceding discussion it should be apparent that any natural number which is the product of two smaller natural numbers is divisible by either of those smaller numbers. In our introductory study of bases we partitioned sets of objects into disjoint subsets. For example, in the following array, 15 asterisks are arranged to show that $15 = 3 \cdot 5$ (or $15 = 5 \cdot 3$) by having

```
5     *  *  *  *  *
5     *  *  *  *  *
5     *  *  *  *  *
──
3 × 5 = 15    3 + 3 + 3 + 3 + 3 = 5 × 3 = 15
```

five asterisks on each of three rows. The same groupings illustrate the facts

$$15 \div 5 = 3 \quad \text{and} \quad 15 \div 3 = 5$$

**Definition 7.2.** For $m \in W$ and $n \in N$, $m \div n$ represents a number $k$ for which $k \cdot n = m$ (if such a number exists); or $m \div n = k$ iff $k \cdot n = m$ where $n \neq 0$. $m \div n$ is read *m divided by n* and is called the *quotient* of $m$ and $n$; $m$ is called the *dividend* and $n$ the *divisor*.

Observe the formal similarity of Definition 7.2 to Definition 7.1. The following statements are similar to the lemmas and to Theorem 7.1. No proofs are given, so that you may test your ability to construct an argument by following a pattern where the argument is similar.

(1) $n \div n = 1$ for every $n \in N$.

(2) $(m \div n) \cdot n = m$ where $m, m \div n \in W$ and $n \in N$.

(3) $(m \cdot n) \div n = m$ where $m \in W$ and $n \in N$.

(4) $h \cdot n = k \cdot n$ implies $h = k$ for $h, k \in W$ and $n \in N$.

*Inverse Operations* **117**

In this discussion, you should have noticed the care taken to insure that $n \neq 0$. The need for this precaution becomes clear when (4) is written with numbers for $h$ and $k$ and with zero for $n$. For example, if (4) is applied incorrectly to the statement

$$3 \cdot 0 = 7 \cdot 0$$

(which is true since $3 \cdot 0 = 0$ and $7 \cdot 0 = 0$ also), what kind of difficulty arises? If Definition 7.2 were relaxed to allow $n = 0$, contradictions like $3 = 7$ would result, and we would no longer have a completely logical structure. It is for this reason that *division by zero is undefined.*

A more common situation which arises when attempts are made to partition sets is illustrated in the following illustration of 17 asterisks "grouped by fives."

$$\begin{array}{l} 5 \quad * \; * \; * \; * \; * \\ 5 \quad * \; * \; * \; * \; * \\ 5 \quad * \; * \; * \; * \; * \\ 2 \quad * \; * \end{array}$$

$$15 + 2 = 3 \times 5 + 2$$

The representation

$$17 = 3 \times 5 + 2$$

recalls the expanded form of the numeral $32_{\text{five}}$ used to represent the number seventeen. As we observed in our study of bases, grouping by fives insures a remainder subset of cardinality *less than* five. This fundamental concept is expressed in a statement known to most mathematicians as the Division Algorithm. Its discovery is attributed to the Greek mathematician Euclid (c. 300 B.C.). We present it here in modified form as a theorem (without proof), the form in which it usually occurs.

## *Theorem 7.4*

For every $b \in W$ and $d \in N$ there are unique $q, r \in W$ such that $b = d \cdot q + r$ and $r < d$.

In the preceding example $b = 17, d = 5, q = 3$ and $r = 2$. When describing this situation to school children we say that 17 is the *dividend*, 5 is the *divisor*, 3 is the *quotient* and 2 is the *remainder*. In case of a zero remainder, as in the preceding example of $15 \div 5$, the algorithm reduces to the simple case described by Definition 7.2. The two examples are clearly distinguished when written according to the division algorithm:

$$17 = 5 \cdot 3 + 2$$

$$15 = 5 \cdot 3 + 0$$

In the following development we make free use of these principles.

# THE DIVISION ALGORITHM

Division has been performed by a number of different algorithms, because of its relation to multiplication. We will confine our study to those most closely related to the principles and processes we have already developed.

In representing division by partitioning sets, we were performing a process which we used in representing numbers in different bases. Division by seven could be performed by finding the "number of sevens" in a set. We will demonstrate a development which proceeds from the simple to the relatively sophisticated. In $35 \div 7$ the number of 7's can be determined by counting by 7's. Thus

$$35 = 7 + 7 + 7 + 7 + 7 = 5 \times 7$$

so that $35 \div 7 = 5$. Another primitive method for determining the number of sevens in 35 is repeated subtraction, the inverse of repeated addition. This is the principle employed by some desk calculators, with the 7's counted as they are subtracted:

```
               No. of Sevens
       35
       -7           1
       ──
       28
       -7           2
       ──
       21
       -7           3
       ──
       14
       -7           4
       ──
        7
       -7           5
       ──
```

All of these processes are slow and wasteful, but they do provide greater insight for pupils who need it. A more economical method is to examine the memorized basic multiplication facts for multiples of 7 which are close to 35.

After practicing the most basic division problems involving single-digit divisors and quotients such as $35 \div 7 = 5$, children are usually directed to related problems of higher place value, such as $350 \div 7$ or $35{,}000 \div 7$. To justify the intuitive procedure of taking the quotient of $35 \div 7$ and multiplying it by 10 or 1,000, we need theorems similar to those developed for multiplication. We would like to be assured that

$$35{,}000 \div 7 = (35 \times 1{,}000) \div 7 = (35 \div 7) \times 1000$$

## Theorem 7.5

If $m \div n \in W$ and $k \in W$, then $(m \times k) \div n = (m \div n) \times k$.

*Proof.* $[(m \div n) \times k] \times n = (m \div n) \times (k \times n) = (m \div n) \times (n \times k) = [(m \div n) \times n] \times k = m \times k$ by the associative and commutative properties of multiplication in $(W, +, \times)$, and from Definition 7.2, which identifies $(m \div n) \times n$ as $m$. Since $[(m \div n) \times k] \times n = m \times k$, by (4) following Definition 7.2, $(m \times k) \div n = (m \div n) \times k$.

EXAMPLE: $5600 \div 8 = (56 \times 100) \div 8 = (56 \div 8) \times 100 = 7 \times 100 = 700$.

EXAMPLE: A more complicated problem such as $5760 \div 8$ can be given a similar treatment if we recognize 5760 as $5600 + 160$ and if our division can be distributed over the sum. Then we could write

$$5760 \div 8 = (5600 + 160) \div 8$$
$$= (5600 \div 8) + (160 \div 8)$$
$$= 700 + 20 = 720$$

## Theorem 7.6

If $m \div k$ and $n \div k$ are defined in $W$, then $(m + n) \div k = (m \div k) + (n \div k)$.

*Proof:* $[(m \div k) + (n \div k)] \times k = [(m \div k) \times k] + [(n \div k) \times k] = m + n$ because multiplication is distributive over addition in $(W; +, \times)$, and by Definition 7.2. Thus by Definition 7.2, $(m \div k) + (n \div k) = (m + n) \div k$.

EXAMPLE: $2608 \div 8$ provides a problem which requires a little more reliance on formal procedure. From our previous example, however, we expect to find a multiple of 8 which is near either 2 or 26 (first digits), and since all non-trivial multiples of 8 are greater than 2, we choose 24 and write

$$2608 = 2400 + 208$$

Next, we seek a multiple of 8 near 20 (for 208), choosing 16 and writing

$$2608 = 2400 + 160 + 48$$

Since 48 is a multiple of 8, we are ready to perform the division:

$$2608 \div 8 = (2400 + 160 + 48) \div 8$$
$$= (2400 \div 8) + (160 \div 8) + (48 \div 8)$$
$$= 300 + 20 + 6$$
$$= 326$$

The preceding work can be accomplished more easily using the following familiar algorithm. The numerals in the left column are shown merely to explain how the numerals to their right were obtained.

$$
\begin{array}{rl}
(\text{quotient}) \longrightarrow & 326 \\
\downarrow (\text{divisor}) \quad 8\,|\, & 2608 \quad (\text{dividend}) \\
300 \times 8 = & 2400 \\ \hline
& 208 \quad (\text{remaining dividend}) \\
20 \times 8 = & 160 \\ \hline
& 48 \quad (\text{remaining dividend}) \\
6 \times 8 = & 48 \\ \hline
\text{Check:} \quad 326 \times 8 + & 0 = 2608
\end{array}
$$

EXAMPLE: One of the advantages of algorithms such as the preceding one is that they make possible the relatively easy solution of complex problems. For example, to calculate $11{,}400 \div 28$ without considering only one or two digits at a time would be extremely difficult. In multiplication, it was convenient to work with the units digits first and progress in an orderly fashion to the digits of higher place value. In the inverse operation of division, it is more convenient to begin with the digits of higher place value, attempting the initial division $1 \div 2$ or $11 \div 2$. Of course, $1 < 2$, so we would choose $11 \div 2$; but 28 is almost 30, so it would be more realistic to *think* $11 \div 3$ as a beginning.

$$
\begin{array}{rl}
(\text{quotient}) & \phantom{00}407 \\
(\text{divisor}) \quad 28\,|\, & 11{,}400 \quad (\text{dividend}) \\
400 \times 28 \quad = & 11{,}200 \\ \hline
& 200 \quad (\text{remaining dividend}) \\
0 \times 28 \quad = & 0 \\ \hline
& 200 \quad (\text{remaining dividend}) \\
7 \times 28 \quad = & 196 \\ \hline
\text{Check:} \quad 407 \times 28 \quad + & 4 = 11{,}400
\end{array}
$$

*Exercise 7.2*

1. Arrange a set of 18 asterisks to illustrate
   (a) $18 \div 6$    (c) $18 \div 3$
   (b) $18 \div 9$    (d) $18 \div 1$

**2.** Partition the set
$$\{(a, x), (b, y), (c, z), (b, x), (a, y), (c, y), (a, z), (b, z), (c, x)\}$$
to illustrate the relation between $9 \div 3 = 3$ and $3 \times 3 = 9$.

**3.** Complete the following:
(a) $14 \div 7 = \underline{\phantom{xx}}$ because $\underline{\phantom{xx}} \times 7 = 14$.
(b) $39 \div \underline{\phantom{xx}} = 3$ because $3 \times \underline{\phantom{xx}} = 39$.
(c) $\underline{\phantom{xx}} \div 17 = 7$ because $7 \times 17 = \underline{\phantom{xx}}$.

**4.** Find the number represented by $n$ if it exists:
(a) $7 \cdot n = 56$     (d) $45 \div n = g$
(b) $n \div 11 = 3$     (e) $n \cdot 0 = 6$
(c) $n \cdot 8 = 0$

**\*5.** Use definitions to express $n$ in terms of the other numbers involved in each equation. (For example: If $10 \div n = 5$ then $5 \times n = 10$, so $n = 10 \div 5$)
(a) $n \cdot 8 = 56$     (d) $n \cdot 0 = 6$
(b) $n \div 4 = 9$     (e) $(n + 3) \cdot 5 = 20$
(c) $n \cdot 6 = 0$

**6.** Calculate the following:
(a) $(96 \div 12) \div 4$     (c) $96 \div (12 \times 4)$
(b) $96 \div (12 \div 4)$     (d) $96 \div 12 \div 4$

**7.** Use Theorems 7.5 and 7.6 to perform the following:
(a) $3400 \div 17$
(b) $(63{,}000 + 2700 + 450) \div 9$
(c) $360 \div 8$

**8.** Use the familiar long-division algorithm to calculate
(a) $66{,}150 \div 90$     (b) $34{,}162 \div 273$

**9.** Write each of the divisions of Problem 8 in the form of Theorem 7.4.

**\*10.** Use definitions and theorems of this section to prove:
(a) $(m \times k) \div (n \times k) = m \div n$ when all indicated numbers exist.
(b) $(m \div n) \div k = m \div (n \times k)$ when all indicated numbers exist. Compare with Problem 6.

**\*11.** Prove the statements (1) through (4) which follow Definition 7.2 on page 116.

## INVERSE OPERATIONS

Although it will be apparent from material in the succeeding chapter that the concept of an operation inverse to another is not essential to the development of algebraic systems, the subtraction and division processes are so well established and practical that they will probably continue to be a part of our mathematics for an indefinite period. Aside from their computational value, they also have a great pedagogical value in that they afford exercise in the formulation of equivalent statements, such as in some of the formal procedures used to find solution sets for equations. The similarity of the definitions for subtraction and division tend to justify the generalization of the following definition, even though there are also some difficulties inherent in the concept.

*Definition 7.3.* A binary operation $*$ on a set $S$ is *inverse* to a binary operation $\bigcirc$ on $S$ iff $(a * b) \bigcirc b = a$ for every $a, b \in S$ for which $a * b \in S$.

In the further development of number systems, we will be concerned with the enlargement of our basic set $S$ to insure its closure with respect to all elementary operations, so that the final qualifying statement of Definition 7.3 will not be necessary. For the present, however, we see that such expressions as

$$12 - 17 \quad \text{and} \quad 8 \div 14$$

are meaningless in our system $(W; +, \times)$ since they do not represent elements of $W$. Let us now turn our attention to exercises in non-decimal bases in order to review both that concept and the more recent one of inverse operations.

## SUBTRACTION IN BASE SIX

We have observed that the usual practice for developing proficiency in subtraction is the memorization of a separate set of *subtraction facts*. Since we do not wish to make this a prerequisite for performing subtraction in non-decimal bases, we will provide the basic *addition table* and adapt it to our purposes by using the definition of subtraction. The base six addition table shown below illustrates the basic relations

$$3_{\text{six}} + 5_{\text{six}} = 12_{\text{six}}; \quad 12_{\text{six}} - 5_{\text{six}} = 3_{\text{six}}; \quad 12_{\text{six}} - 3_{\text{six}} = 5_{\text{six}}$$

| + | 0 | 1 | 2 | 3 | 4 | 5 |
|---|---|---|---|---|---|---|
| 0 | 0 | 1 | 2 | 3 | 4 | 5 |
| 1 | 1 | 2 | 3 | 4 | 5 | 10 |
| 2 | 2 | 3 | 4 | 5 | 10 | 11 |
| →3 | 3 | 4 | 5 | 10 | 11 | (12) |
| 4 | 4 | 5 | 10 | 11 | 12 | 13 |
| 5 | 5 | 10 | 11 | 12 | 13 | 14 |

The digit-by-digit procedure of the following examples makes use of the table in this fashion.

EXAMPLE:

| Short Form | Semi-expanded | Expanded Form |
|---|---|---|

$$435_{six} = 400_{six} + 30_{six} + 5_{six} = 4 \times 100_{six} + 3 \times 10_{six} + 5$$
$$-231_{six} = 200_{six} + 30_{six} + 1_{six} = 2 \times 100_{six} + 3 \times 10_{six} + 1$$

$$204_{six} = 200_{six} \phantom{+ 30_{six}} + 4_{six} = 2 \times 100_{six} + 0 \times 10_{six} + 4$$

EXAMPLE:

$$\begin{array}{c}4\phantom{0}12\phantom{0}12\\\cancel{5}\phantom{0}\cancel{3}\phantom{0}\cancel{2}_{six}\end{array} = 400_{six} + 120_{six} + 12_{six} = 4 \times 100_{six} + 12_{six} \times 10_{six} + 12_{six}$$
$$-2\phantom{00}3\phantom{00}4_{six} = 200_{six} + \phantom{0}30_{six} + \phantom{0}4_{six} = 2 \times 100_{six} + \phantom{0}3\phantom{0} \times 10_{six} + \phantom{0}4_{six}$$

$$\phantom{-}2\phantom{00}5\phantom{00}4_{six} = 200_{six} + \phantom{0}50_{six} + \phantom{0}4_{six} = 2 \times 100_{six} + \phantom{0}5\phantom{0} \times 10_{six} + \phantom{0}4_{six}$$

In the last example, the change in form can be demonstrated in steps:

$$532_{six} = 5 \times 100_{six} + 3 \times 10_{six} + 2$$

$$= 5 \times 100_{six} + 2 \times 10_{six} + 1 \times 10_{six} + 2$$

$$= 5 \times 100_{six} + 2 \times 10_{six} + 12_{six}$$

$$= 4 \times 100_{six} + 1 \times 100_{six} + 2 \times 10_{six} + 12_{six}$$

$$= 4 \times 100_{six} + 10_{six} \times 10_{six} + 2 \times 10_{six} + 12_{six}$$

$$= 4 \times 100_{six} + (10_{six} + 2) \times 10_{six} + 12_{six}$$

$$= 4 \times 100_{six} + 12_{six} \times 10_{six} + 12_{six}$$

## OTHER FORMS OF SUBTRACTION

We have already investigated the use of the number line to illustrate subtraction, where the subtraction of a whole number is accomplished by counting backward the stated number of units. Since "clock addition" is performed in a manner similar to addition on the number line, it would seem that "clock subtraction" should consist of counting backward on the face of the clock. The following figures picture an example of the inverse relation of these operations. A table of basic "clock addition" facts can be compiled, from which you can read subtraction facts in the manner demonstrated on the base six addition table.

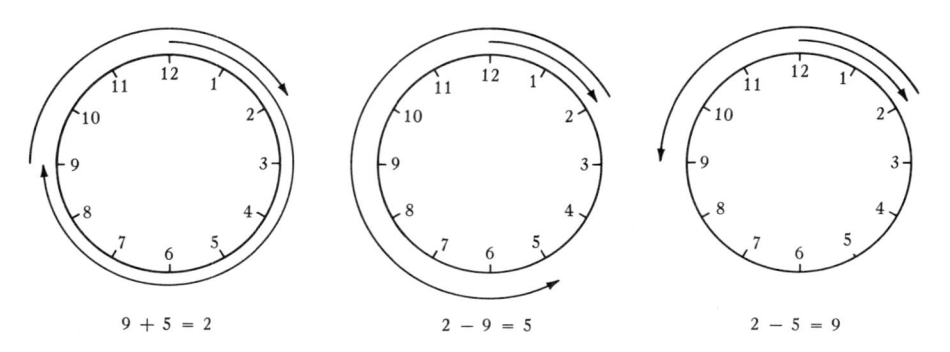

$$9 + 5 = 2 \qquad 2 - 9 = 5 \qquad 2 - 5 = 9$$

A number of algorithms for performing subtraction are known, most of which have only curiosity value. The following example is furnished to give the student an opportunity to analyze what *true* mathematical processes were used and from this to verify that the procedure is indeed valid.

EXAMPLE. Calculate $723 - 456$.

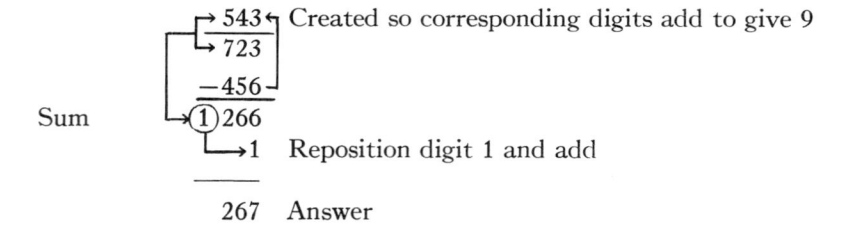

```
              ┌→ 543 ┐   Created so corresponding digits add to give 9
              └→ 723 │
                -456 ┘
Sum           └→ ①266
               └→1      Reposition digit 1 and add
              ─────
                267     Answer
```

At first glance, this method seems to accomplish subtraction without *performing* any subtraction. However, the digits of the numeral 543 were obtained by subtracting 456 from 999.

## DIVISION IN BASE SIX

The inverse relation between multiplication and division allows the use of the table of base six multiplication facts as a basis for performing division in base six. The numerals marked in the table below illustrate the related basic facts,

$$4_{six} \times 5_{six} = 32_{six}; \quad 32_{six} \div 4_{six} = 5_{six}; \quad 32_{six} \div 5_{six} = 4_{six}$$

| × | 0 | 1 | 2 | 3 | 4 | ↓5 |
|---|---|---|---|---|---|---|
| 0 | 0 | 0 | 0 | 0 | 0 | 0 |
| 1 | 0 | 1 | 2 | 3 | 4 | 5 |
| 2 | 0 | 2 | 4 | 10 | 12 | 14 |
| 3 | 0 | 3 | 10 | 13 | 20 | 23 |
| →4 | 0 | 4 | 12 | 20 | 24 | (32) |
| 5 | 0 | 5 | 14 | 23 | 32 | 41 |

Since place value numeration takes on the same form in every base, division in base six can be performed by the same algorithm as decimal division. In the following examples, each digit of the quotient is selected by finding the "nearest" one- or two-digit product in the table equal to or less than the corresponding one- or two-digit numeral in the dividend.

EXAMPLE: In examining $212_{six} \div 4_{six}$ we observe that $4 > 2$, so we look in the table for a multiple of 4 which is equal to or less than 21. Since this turns out to be 20, we rewrite $212_{six}$ as $200_{six} + 12_{six}$. Then

$$212_{six} \div 4_{six} = (200_{six} + 12_{six}) \div 4_{six}$$
$$= (20_{six} \div 4_{six}) \times 10_{six} + 12_{six} \div 4_{six}$$
$$= 3 \times 10_{six} + 2_{six}$$
$$= 32_{six}$$

where 3 and 2 were found from the table. In the conventional form, the algorithm is as follows:

$$\begin{array}{r} 32_{six} \\ 4_{six}\overline{)212_{six}} \\ 200 \\ \hline 12 \\ 12 \\ \hline \end{array}$$

126                Elementary Mathematical Structure

EXAMPLE: An example which affords the complications of a "trial divisor" digit and a remainder is $3125_{six} \div 43_{six}$. Here the multiplication becomes more involved, and the addition table will be desirable for performing some of the subtraction.

$$\begin{array}{r} 41_{six} \\ 43_{six} \overline{)3125_{six}} \\ 3000 \\ \hline 125 \\ 43 \\ \hline 42_{six} \end{array}$$

## Exercise 7.3

1. Let $A = \{a, b, c, d, e, f, g, h, i, j, k, m\}$ and $B = \{x, y, z\}$. Try to devise a set operation using $A$ and $B$ to produce a set $C$ for which $n(A) \div n(B) = n(C)$.

2. Use a numerical example and Definition 7.3 to illustrate each of the following statements:
   (a) Addition on $S$ is inverse to subtraction on $S$.
   (b) Multiplication on $S$ is inverse to division on $S$.

3. Use the base six addition table as needed to fill in the blanks in the following:
   (a) $4_{six} + 5_{six} = 13_{six}$; therefore, $13_{six} - 5_{six} = $ ____.
   (b) $12_{six} - 3_{six} = $ ____ because ____ $+ 3_{six} = 12_{six}$.
   (c) $41_{six} - 23_{six} = $ ____.
   (d) $4352_{six} - 1254_{six} = $ ____.

4. Perform the following subtractions in the indicated bases. Check your results by expressing all numbers in base ten.
   (a) $56_{eight} - 23_{eight} = $ ____.
   (b) $1110_{two} - 101_{two} = $ ____.

5. Perform each of the following in both expanded form and modified expanded form.
   (a) $345_{six} - 132_{six}$   (b) $432_{six} - 145_{six}$

6. In clock arithmetic, $2 \times 9 = 6$; therefore,
   (a) $6 \div 2 = $ ____, and  (b) $6 \div 9 = $ ____.

7. We have seen that division can be considered repeated subtraction. For example, $28 - 7 - 7 - 7 - 7 = 0$ so that $28 \div 7 = 4$. Use this principle to demonstrate the following.
   (a) $15 \div 3$   (b) $18 \div 6$

8. Perform the following divisions on the face of a clock, by repeated subtraction:
   (a) $10 \div 2$   (b) $6 \div 9$   (c) $6 \div 2$

   Compare with Problem 6(a). Do you think the results of division in clock arithmetic are always unique?

9. Fill in the blanks to make true statements or explain why the operation is not possible:
   (a) $0 \div 3 =$ ____ because ____ $\times 3 = 0$.
   (b) $3 \div 0 =$ ____ because ____ $\times 0 = 3$.
   (c) $0 \div 0 =$ ____ because ____ $\times 0 = 0$.

10. Let $(S; \bigcirc, *)$ be defined by $S = \{a, b, c\}$ and by the addition and multiplication tables shown.

    addition

    | $\bigcirc$ | a | b | c |
    |---|---|---|---|
    | a | a | b | c |
    | b | b | c | a |
    | c | c | a | b |

    multiplication

    | $*$ | a | b | c |
    |---|---|---|---|
    | a | a | a | a |
    | b | a | b | c |
    | c | a | c | b |

    Calculate the following from the tables where $-$ is the inverse of $\bigcirc$ and $\div$ is the inverse of $*$.
    (a) $a - c$
    (b) $c - b$
    (c) $b \div c$
    (d) $(c - b) \div (b - c)$
    (e) Is $b * (a \bigcirc c) = (b * a) \bigcirc (b * c)$? Show your work.

11. Use the base six multiplication table and a base six addition table to perform the following divisions:
    (a) $1422_{six} \div 34_{six}$
    (b) $35300_{six} \div 43_{six}$
    (c) $15304_{six} \div 312_{six}$

# 8

# Extending Number Systems

Following our practice of increasing abstraction and diminishing reliance on concrete models established in Chapter 7, we will continue to use the concept of inverse relations in the development. The basic technique of this particular development is to select two elements $b$ and $c$ at random from a set $S$ for which an operation $*$ is defined, and ask the question, "For what element $x$ is $b * x = c$?" Such questions are increasingly prominent in mathematics textbooks for the elementary grades, and take such forms as

$$6 + \_\_ = 13$$
$$5 + \_\_ < 9$$

or

$$11 - 2n > 3$$

depending upon the grade level. The student is expected to supply a list of all elements from a designated set which make true statements when inserted in place of the blank, or letter. These elements are called *solutions of the open sentences*, and the set of solutions of an open sentence is called a *solution set* of that sentence. We will use these ideas often, so it will be helpful to have formal definitions. Just as we used $p$ and $q$ to represent statements in our discussion of Language and Mathematics, we shall use $p(x)$ to represent an open sentence concerning something represented by $x$. For example, $p(x)$ could represent the sentence

$$6 + x = 13$$

*Definition 8.1.* An *open sentence in x*, represented by $p(x)$, is a declarative sentence concerning $x$. The letter $s$ represents a *solution* of $p(x)$ iff $p(s)$ is a true statement. For any set $S$, the *solution set of $p(x)$ relative to $S$* is the set $\{y \mid y \in S \text{ and } p(y) \text{ is true}\}$. If the verb of $p(x)$ is *equals*, then $p(x)$ is called an *equation*. $p(x)$ and $q(x)$ are *equivalent* iff they have the same solution set.

EXAMPLE: $6 + x = 13$ is an equation whose only solution is 7.

EXAMPLE: The solution set of $11 - 2x > 3$ relative to the set $W$ is $\{0, 1, 2, 3\}$, since the use of these elements in place of $x$ yields the statements
$$11 > 3, \quad 9 > 3, \quad 7 > 3, \text{ and } 5 > 3$$
respectively, all of which are true, while the use of any other whole number does not. For example, if 4 replaces $x$, the statement is $3 > 3$, which is false, while the use of 6 produces $11 - 12 > 3$ which cannot presently be classified as a statement since $11 - 12$ is undefined in our development.

## THE INTEGERS

In our attempt to find all elements of $W$ which are solutions of
$$11 - 2x > 3$$
we find ourselves confronted with a situation which we have carefully avoided until now. Do expressions such as
$$11 - 12$$
or
$$0 - 7$$
represent numbers? From a more realistic point of view (since *number* is a product of the mind) a better question would be, "*Can* there be numbers of this type?"

From this viewpoint, our problem is to invent more numbers, provided that we wish the relation between addition and subtraction given by Definitions 7.1 and 7.3 to stand without qualification. Such a generalized form of Definition 7.1 would demand that $11 - 12$ represent a number for which
$$(11 - 12) + 12 = 11$$
But the properties of $(W; +, \times)$ demand that
$$m + 12 \geq 12$$
or
$$m + 12 > 11$$
for every $m \in W$, so that $11 - 12$ would have to represent some new kind of number.

The process of inventing new numbers or elements, or generalizing operations and relations of a system to create a more comprehensive system, is called *extending* the system. $(W; +)$ was extended to $(W; +, \times)$ by introducing a new operation. The system $(W; +)$ is an *extension* of the system

$(N; +)$ because of the extra element zero. Since an extension of a system must not deny the original properties of its operations, at least for its old elements, it seems desirable to try to generalize all operations, relations and the basic set so as to preserve these properties. Let us see how we will define the set

$$\{x \mid x + b = c \quad \text{for} \quad b, c \in W\}$$

and the operations $+$ and $\times$ in order that $+$ and $\times$ still have the associative, commutative and distributive properties.

EXAMPLE: If we assume associativity of addition for the system in which the equation

$$x + 13 = 6$$

occurs, then an equivalent equation is

$$(x + 7) + 6 = 6$$

Furthermore, since the additive identity is unique (Exercise 6.1, Problem 10)

$$x + 7 = 0$$

is also equivalent to the first equation.

From the example, we can observe that when $b > c$, there is associated with the equation

$$x + b = c$$

an equivalent equation

$$x + (b - c) = 0$$

where $b - c \in N$. Therefore, the new numbers we are seeking belong to the set

$$\{y \mid y + n = 0 \quad \text{for} \quad n \in N\}$$

and it is desirable to have one such number to associate with each natural number.

## Axiom 8.1

For every $n \in N$ there is a unique number represented by $^-n$ called *negative n* and a binary operation $\oplus$ called *addition* for which $n \oplus {^-n} = 0$.

*Definition 8.2.* The elements of the set $J = \{\cdots, ^-2, ^-1, 0, 1, 2, \cdots\}$ are called *integers*. The set $\{^-1, ^-2, ^-3, \cdots\}$ is the set of *negative integers*, and elements of $N = \{1, 2, 3, \cdots\}$ are also called *positive integers*.

## THE SYSTEM $(J; \oplus, \otimes)$

In order to investigate the nature of the new system of integers with new addition and multiplication operations, we must specify our basic assumptions about the operations.

## Axiom 8.2

There is a binary operation $\otimes$ called multiplication on $J$. For all $m, n \in W$, $m \oplus n = m + n$ and $m \otimes n = m \times n$. $\oplus$ and $\otimes$ are associative and commutative, and $\otimes$ is distributive over $\oplus$.

EXAMPLE:
$$\begin{aligned}
{}^-7 \oplus 13 &= {}^-7 \oplus (7 \oplus 6) \\
&= ({}^-7 \oplus 7) \oplus 6 \\
&= 0 + 6 \qquad [\text{System } (W; +, \times)] \\
&= 6
\end{aligned}$$

In this example the problem of addition in the case of $^-7 \oplus 7$ was handled by Axiom 8.1, while that of $0 \oplus 6$ was an application of Axiom 8.2. What would happen in the case $0 + {}^-6$?

EXAMPLE:
$$\begin{aligned}
0 &= 0 + 0 \\
&= 0 \oplus ({}^-6 \oplus 6) \\
&= (0 \oplus {}^-6) \oplus 6
\end{aligned}$$

From Axiom 8.1 the statement that

$$0 = x \oplus 6$$

identifies $X$ uniquely as $^-6$, so that our investigation in this example shows that

$$0 \oplus {}^-6 = {}^-6.$$

It should also be apparent that the principle would apply not only to $^-6$ but to every $^-n$ where $n \in N$; that is

$$0 \oplus {}^-n = {}^-n \quad \text{for every} \quad n \in N$$

Thus, zero is an additive identity for every negative integer. Since it also has this property for the whole numbers, we have the following:

## Theorem 8.1

Zero is the identity for $\oplus$ in $(J; \oplus, \otimes)$.

EXAMPLE:
$$\begin{aligned}
{}^-13 \oplus 7 &= {}^-13 \oplus 7 \oplus 0 \\
&= {}^-13 \oplus 7 \oplus (6 \oplus {}^-6) \\
&= [{}^-13 \oplus (7 \oplus 6)] \oplus {}^-6 \\
&= ({}^-13 \oplus 13) \oplus {}^-6 \\
&= 0 \oplus {}^-6 \\
&= {}^-6
\end{aligned}$$

## Extending Number Systems

EXAMPLE: $^-6 \oplus {}^-7 = {}^-6 \oplus {}^-7 \oplus 0$
$\qquad = {}^-6 \oplus {}^-7 \oplus (13 \oplus {}^-13)$
$\qquad = ({}^-6 \oplus {}^-7 \oplus 13) \oplus {}^-13$
$\qquad = ({}^-6 \oplus {}^-7 \oplus 6 \oplus 7) \oplus {}^-13$
$\qquad = ({}^-6 \oplus 6) \oplus ({}^-7 \oplus 7) \oplus {}^-13$
$\qquad = 0 \oplus 0 \oplus {}^-13$
$\qquad = {}^-13$

The emphasis in these examples has been that addition as described for $(J; \oplus, \otimes)$ is meaningful and can be performed simply by using the basic properties of associativity, commutativity, and zero's additive identity property. Before presenting a set of exercises, we shall demonstrate the same examples on the extended number line of the integers.

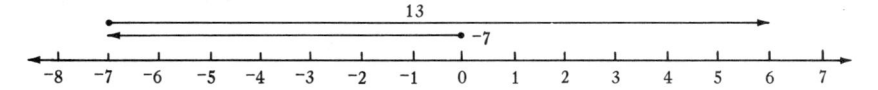

EXAMPLE: $^-7 + 13 = 6.$

EXAMPLE: $7 + {}^-13 = {}^-6.$

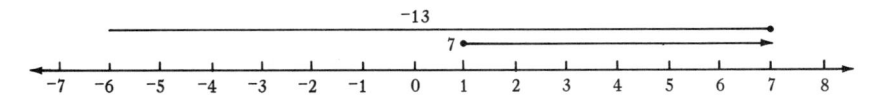

EXAMPLE: $^-6 + {}^-7 = {}^-13.$

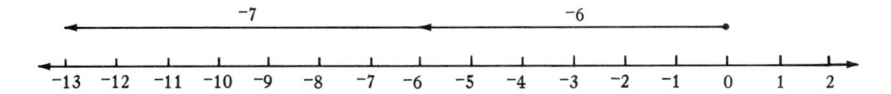

In the preceding development we have introduced new symbols for addition and multiplication to emphasize the fact that these are new operations. However, since it is somewhat of a chore to formulate and remember to use these unfamiliar symbols, and since everyone uses the standard symbols $+$ and $\cdot$ anyway, we shall adopt these traditional symbols for the basic operations, writing $(J; +, \times)$ for the system. This decision will allow you greater concentration on the ideas in the following exercises by demanding less attention to unaccustomed notation.

*Exercise 8.1*

1. Find solutions for the following equations:
   (a) $x + 13 = 25$
   (b) $x + 9 = 9$
   (c) $x = 9 - 9$
   (d) $x = 9 + {}^-9$
   (e) $y + {}^-7 = 0$

*2. Write the solution set for the following:
   (a) $y = (2 + 3) + {}^-3$
   (b) $2 \cdot x = 5 + {}^-3$
   (c) $(y + 1) + {}^-7 = 0$
   (d) $3 \cdot x - 5 = 7$
   (e) $3(y - 2) + {}^-18 = 0$

3. Identify the elements of the following sets:
   (a) $\{x \mid x \in W, 5x < 20\}$
   (b) $\{y \mid 2 \leq y + 1 < 3, y \in W\}$
   (c) $\{x \mid x \in N, 21 < 7x < 27\}$

*4. In view of our present development, is 5 a solution of either of the following equations? Justify your conclusions.
   (a) $(x + {}^-5) \cdot (x + 2) = 0$
   (b) $(x + {}^-5) \cdot (x + {}^-7) = 0$

5. Which of the following open sentences is equivalent to $15 - 2 \cdot n = 3$ where $n \in N$?
   (a) $15 = 3 + 2 \cdot n$
   (b) $15 - 3 = 2n$
   (c) $2 \cdot n - 15 = 3$
   (d) $n = (15 - 3) \div 2$
   (e) $n + 2 = 8$

*6. Use the methods of the examples to calculate the following:
   (a) ${}^-8 + {}^-9$
   (b) ${}^-11 + 4$
   (c) $5 + {}^-3$
   (d) ${}^-1 + {}^-4$
   (e) $17 + {}^-6$
   (f) $10 + {}^-19$

7. Demonstrate the following on number lines:
   (a) ${}^-4 + {}^-5$
   (b) $6 + {}^-5$
   (c) ${}^-5 + 6$
   (d) ${}^-6 + 5$

*8. The third example in this section shows that $7 + {}^-13 = {}^-6 = {}^-(13 - 7)$. Use the natural numbers associated with the integers of each part of Problem 6 to express the sums similarly.

9. Find the solution in $J$ of the following equations:
   (a) $x + {}^-7 = 0$
   (b) $(y + {}^-2) + {}^-3 = {}^-5$
   (c) $x + 12 = 4$
   (d) $n + n = {}^-10$
   (e) ${}^-6 + n = {}^-17$
   (f) ${}^-(n + 1) + 5 = 0$

*10. If for $x, y \in J$ we define $x - y$ as a number for which $(x - y) + y = x$ and ${}^-y$ as a number for which $y + {}^-y = 0$, show that $x - y = x + {}^-y$.

# OPERATIONS IN $(J; \oplus, \otimes)$

In this section we shall generalize and summarize our findings concerning addition in $(J; \oplus)$ and then examine multiplication in $(J; \oplus, \otimes)$. The following theorem shows how addition of integers could be defined in terms of addition and subtraction of whole numbers. For the purpose of simplifying the statement and proof of the theorem we shall allow $^-0 = 0$, so that $^-0 + 0 = 0 + {}^-0 = 0$ also follows.

## *Theorem 8.2*

(1) If $a, b \in W$ then $a \oplus b = a + b$ and $^-a \oplus {}^-b = {}^-(a + b)$.
(2) If $a, b \in W$ with $a \geq b$ then $a \oplus {}^-b = a - b$ and $^-a \oplus b = {}^-(a - b)$.

*Proof:* (1) The first case is simply a restatement of part of Axiom 8.2. $^-a \oplus {}^-b = ({}^-a \oplus {}^-b) \oplus [(a + b) \oplus {}^-(a + b)] = [({}^-a \oplus {}^-b) \oplus (a \oplus b)] \oplus {}^-(a + b) = ({}^-a \oplus a) \oplus ({}^-b \oplus b) \oplus {}^-(a + b) = 0 \oplus 0 \oplus {}^-(a + b) = {}^-(a + b)$. (2) If $a, b \in W$ with $a \geq b$, then $a - b \in W$ by Definitions 2.5 and 7.1. Then $a \oplus {}^-b = [(a - b) + b] \oplus {}^-b = (a - b) \oplus b \oplus {}^-b = (a - b) + (b \oplus {}^-b) = (a - b) + 0 = a - b$, and $^-a \oplus b = {}^-[(a - b) + b] \oplus b = [{}^-(a - b) \oplus {}^-b] \oplus b$ [using part (1)] which can be written $^-(a - b) \oplus ({}^-b \oplus b) = {}^-(a - b) \oplus 0 = {}^-(a - b)$.

Observe that the methods of proof used in Theorem 8.2 are the same as those used in the preceding exercises and examples. You should see the results of Theorem 8.2 as a natural consequence; do not use the theorem as a definition.

As we turn our attention to the task of analyzing multiplication of integers, we first remind ourselves that

$$a \otimes b = a \times b$$

for all whole numbers $a$ and $b$. Therefore, we need examine only products where at least one factor is a negative integer. Suppose $n \in N$ so that $^-n$ is a negative integer, and let $b$ represent any whole number. Then $b \times n = bn = nb$ is also a whole number, and

$$^-n \otimes b = {}^-n \otimes b \oplus [(bn) \oplus {}^-(bn)]$$
$$= (b \otimes {}^-n \oplus b \otimes n) \oplus {}^-(bn)$$
$$= b \otimes ({}^-n \oplus n) \oplus {}^-(nb)$$
$$= b \times 0 \oplus {}^-(n \times b)$$
$$= 0 \oplus {}^-(n \times b)$$
$$= {}^-(n \times b)$$

You should observe the use of the fact that $\otimes$ is distributive over $\oplus$ as provided by Axiom 8.2. Also, by allowing $^-0 = 0$, notice that

$$^-n \otimes 0 = {}^-(n \times 0) = {}^-0 = 0$$

so that the multiplicative property of zero is retained in $(J; \oplus, \otimes)$. This property plays an important role when we use the same technique to investigate the product of two negative integers.

If $m, n \in N$ then $^-m$ and $^-n$ are both negative integers and the properties of $(J; \oplus, \otimes)$ allow us to write the following:

$$\begin{aligned}
{}^-m \otimes {}^-n &= {}^-m \otimes {}^-n \oplus [{}^-(mn) \oplus (mn)] \\
&= ({}^-m \otimes {}^-n \oplus {}^-m \otimes n) \oplus (mn) \\
&= {}^-m \otimes ({}^-n \oplus n) \oplus (m \times n) \\
&= {}^-m \otimes 0 \oplus (m \times n) \\
&= 0 + (m \times n) \\
&= m \times n
\end{aligned}$$

It now seems possible to summarize the cases of multiplication in $(J; \oplus, \otimes)$ as we did for addition. Again the statement of the theorem can be simplified if we notice that the last result holds even for $m, n \in W$. For example,

$$^-m \otimes {}^-0 = {}^-m \otimes 0 = 0 = {}^-0 = {}^-(m \times 0)$$

and $^-0 \times {}^-n = {}^-n \times {}^-0$ is reduced to the same type of result.

## Theorem 8.3

(1) If $m, n \in W$ then $m \otimes n = m \times n$ and $^-m \otimes {}^-n = m \times n$.

(2) If $m, n \in W$ then $^-m \otimes n = n \otimes {}^-m = {}^-(m \times n)$.

## INVERSE OPERATIONS IN $(J; +, \times)$

Having accomplished our purpose of describing the operations of $(J; \oplus, \otimes)$ in terms of the system $(W; +, \times)$ we now drop the use of the special symbols $\oplus$ and $\otimes$, expecting that you will recognize the operations represented by $+$ and $\times$ by the context and the set of numbers used in the system. The success achieved in extending the concepts of the two basic operations of $(W; +, \times)$ should encourage a similar attempt for subtraction and division. You may have noticed that some of the problems of Exercise 8.1 were slanted in this direction. Problem 10 of that group shows how mathematicians obviate the necessity of learning a new set of basic operation facts by using a related operation and a related number instead. For example, in working

```
         ←————————————————•
              4 •————————————→
    |————|————|————|————|————|————|————|————|————|
    ⁻3   ⁻2   ⁻1    0    1    2    3    4    5
                    4 − 6 = ⁻2
```

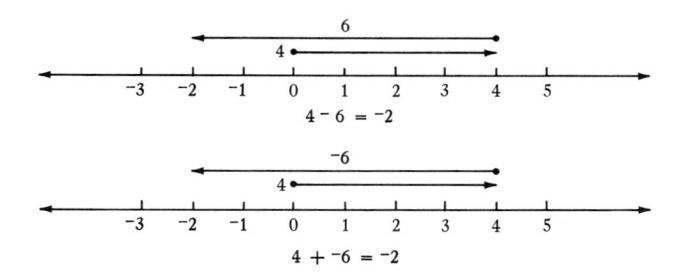

```
              ←——————————————
                 ⁻6
              4 •————————————→
    |————|————|————|————|————|————|————|————|————|
    ⁻3   ⁻2   ⁻1    0    1    2    3    4    5
                   4 + ⁻6 = ⁻2
```

with the number line, you probably observed that the operation of subtracting six is equivalent to that of adding negative six. Just as we described the relation between addition and subtraction by saying that each operation was inverse to the other, we can also say that the numbers 6 and ⁻6 are inverses of each other with respect to addition.

*Definition 8.3.* $b \in S$ is an *inverse* of $c \in S$ with respect to an operation $*$ iff $b * c = c * b = e$ where $e$ is the identity element of $*$ in $(S; *)$. $^-c$ usually denotes an inverse of $c$ with respect to addition, while $c^{-1}$ represents a multiplicative inverse of $c$.

Perhaps you are relieved that we have finally given a name to this relationship between pairs of integers. Each integer has a unique additive inverse shown by the pairings

$$(0, 0), (1, ^-1), (^-2, 2), (^-3, 3), (4, ^-4), \cdots$$

which indicate that zero is its own inverse (remember $^-0 = 0$?). Although some people like to refer to these pairs as *opposites* (perhaps because of their relative positions on the number line) we shall speak of two numbers in a pair as *negatives* of each other. For example, the negative of $^-7$ is 7, or in symbols,

$$-(^-7) = 7$$

According to Definition 7.3, to extend subtraction and division to $(J; +, \times)$, we need only identify these operations in $(J; +, \times)$ as being inverse to $+$ and $\times$ respectively. In order to decide whether such an extension is meaningful, we will examine

$$^-7 - 5$$

If $^-7 - 5$ represents an integer, then by Definition 7.3,

$$(^-7 - 5) + 5 = ^-7$$

Thus we need to solve the equation

$$x + 5 = ^-7$$

138     *Elementary Mathematical Structure*

A moment's reflection suggests that
$$x = {}^-12$$
or
$${}^-7 - 5 = {}^-12$$
for where 12 is related to 7 and 5 in our experience, we are encouraged to generalize. If $x$ and $y$ represent integers for which $x - y$ is also an integer, then ${}^-y \in J$ by Definition 8.3 and previous observations, so that
$$x - y = (x - y) + (y + {}^-y)$$
$$= [(x - y) + y] + {}^-y$$
$$= x + {}^-y$$
Since Theorem 8.2 shows that $J$ is closed under $+$, and since $x$ and ${}^-y$ are integers, the preceding argument shows that $x - y$ represents an integer $x + {}^-y$ because $(x + {}^-y) + y = x + ({}^-y + y) = x + 0 = x$, satisfying Definition 7.3.

## Theorem 8.4

For all $x, y \in J$, $x - y = x + {}^-y$.

EXAMPLE: By Theorem 8.4, the example ${}^-7 - 5$ becomes ${}^-7 + {}^-5 = {}^-12$, agreeing with the observed solution. Similarly,
$${}^-3 - {}^-8 = {}^-3 + {}^-({}^-8) = {}^-3 + 8 = 5$$
This principle of using addition to perform subtraction exemplifies the need to develop the ability to follow abstraction confidently, no matter how far removed from intuitive connections it may become. Of course, the logical connections do exist, and you should thoroughly understand them, in order to acquire this confidence. In other words, convince yourself to the point of absolute certainty that it is quite natural and logical to add the negative of an integer in order to subtract the integer.

By Theorem 8.4, we found that $x - y$ is an integer for all integers $x$ and $y$ essentially because, for every pair of whole numbers $m$ and $n$, $m + n$ and either $m - n$ or $n - m$ are also whole numbers. Since it may happen that neither $m \div n$ nor $n \div m$ is a whole number, we do not expect $x \div y$ to always be an integer. Since Definition 7.3 does not demand this condition for classifying division as inverse to $x$ in $(J; +, \times)$, we will briefly examine typical permissible cases. Since
$$6 \div 3 = 2$$
is defined, we may expect some success by using ${}^-6$ and ${}^-3$ in various combinations.

$$({}^-6 \div 3) \times 3 = {}^-6 \quad \text{implies} \quad {}^-6 \div 3 = {}^-2 = {}^-(6 \div 3)$$
$$(6 \div {}^-3) \times {}^-3 = 6 \quad \text{implies} \quad 6 \div {}^-3 = {}^-2 = {}^-(6 \div 3)$$
$$({}^-6 \div {}^-3) \times {}^-3 = {}^-6 \quad \text{implies} \quad {}^-6 \div {}^-3 = 2 = 6 \div 3$$

# Extending Number Systems

These cases should make the following theorem seem evident and its proof a natural consequence of similar arguments.

## Theorem 8.5

For all $m, n \in W$ for which $m \div n \in W$, (1) $^-m \div {}^-n = m \div n$ and (2) $^-m \div n = {}^-(m \div n) = m \div {}^-n$.

## Exercise 8.2

1. Use Theorem 8.2 to calculate the following, writing details as follows: $^-3 + {}^-2 = {}^-(3+2) = {}^-5$.
   (a) $^-8 + {}^-9$
   (b) $^-11 + 4$
   (c) $5 + {}^-3$
   (d) $^-1 + {}^-4$
   (e) $17 + {}^-6$
   (f) $10 + {}^-19$

2. Use Theorem 8.3 to calculate the following, writing details as follows: $3 \times {}^-2 = {}^-(3 \times 2) = {}^-6$.
   (a) $^-6 \times {}^-5$
   (b) $5({}^-6)$
   (c) $^-7 \times 0$
   (d) $({}^-8) \cdot 4$
   (e) $({}^-1)({}^-10)$
   (f) $8({}^-1)$

3. Verify part (2) of Theorem 8.3 in the following products by using the "repeated addition" concept of multiplication. For example, $3 \cdot ({}^-2) = {}^-2 + {}^-2 + {}^-2 = {}^-4 + {}^-2 = {}^-6 = {}^-(3 \times 2)$.
   (a) $2 \times {}^-6$
   (b) $4({}^-5)$
   (c) $^-3 \times 4$

4. Give the negative of each of the following:
   (a) 6
   (b) 0
   (c) $^-8$
   (d) $^-14 + 9$
   (e) $^-2 \times 3$

*5. Show that inverses are unique; that is, if $b$ and $d$ are both inverses of $c$ in $(S; *)$ then $b = d$. Use $b * c = c * b = e$, $d * c = c * d = e$, $b * e = b$ and $e * d = d$ where $e$ is the identity element for $*$ in $(S; *)$ and $*$ is associative.

6. In clock addition, 12 is the identity element and $^-7 = 5$; that is, 5 is the additive inverse of 7 because $7 + 5 = 12$. Find each of the following:
   (a) $^-3$
   (b) $^-6$
   (c) $^-10$
   (d) $^-4$

7. Use Theorem 8.4 to calculate the following, writing details in the form $^-8 - {}^-4 = {}^-8 + {}^-({}^-4) = {}^-8 + 4 = {}^-(8-4) = {}^-4$.
   (a) $11 - {}^-3$
   (b) $^-3 - 11$
   (c) $0 - 6$
   (d) $^-4 - 0$
   (e) $^-6 - {}^-5$
   (f) $^-9 - {}^-9$

140    Elementary Mathematical Structure

8. See if Theorem 8.4 will adapt to clock subtraction by using the results of Problem 6. For example, $2 - 5 = 2 + {}^-5 = 2 + 7 = 9$ which checks with the example of Chapter 7.
   (a) $7 - 3$   (c) $3 - 10$
   (b) $2 - 4$   (d) $4 - 4$

9. Use Theorem 8.5 to calculate the following, in the form ${}^-10 \div 2 = {}^-(10 \div 2) = {}^-5$.
   (a) $12 \div {}^-4$   (c) ${}^-20 \div 4$
   (b) ${}^-18 \div {}^-3$   (d) $0 \div {}^-6$

*10. Show that the Division Algorithm can be adapted to the integers in the following problems; that is, for integers $b$ and $d \neq 0$ there are unique integers $q$ and $r$ such that $b = d \cdot q + r$ and $0 < r < |d|$, where $|d|$ represents $d$ if $d$ is positive or ${}^-d$ if $d$ is negative. For ${}^-213$ and ${}^-17$, $q = 13$ and $r = 8$ since ${}^-213 = {}^-17(13) + 8$ and $0 < 8 < 17$.
   (a) $320$ and ${}^-23$   (b) ${}^-410$ and $19$   (c) ${}^-267$ and ${}^-13$

*11. Prove that 1 is the multiplicative identity for the system $(J; +, \times)$.

*12. Prove for $m, n \in N$ that ${}^-m \otimes n = {}^-n \otimes m$.

*13. Check the results of Problem 9 by Definition 7.3.

## INVERSES FOR MULTIPLICATION

We have seen that the system $(J; +, \times)$ is an extension of $(W; +, \times)$ essentially through the concept of subtraction as an operation inverse to addition. It has been traditional in the elementary grades to extend the system $(W; +, \times)$ first through division, so that the study of common fractions preceded work with negative numbers. We have selected a development of number systems which, it is hoped, makes it apparent that there are such alternatives. At the same time, a teacher should be acquainted with the order of development indicated by the mathematician's classification of sets of numbers. So, we will extend the system $(J; +, \times)$ by inventing multiplicative inverses for the elements of $J$ just as the negative integers were introduced as additive inverses for the elements of $N$. As before, along with the problem of postulating new numbers, we are also faced with the need to generalize the concepts of addition and multiplication. Again, we will use special symbols for these operations in the new system, to emphasize our knowledge of $(J; +, \times)$ in exploring the nature of the new system. Since inverse elements are directly related to inverse operations and since division by zero is undefined, we should not expect zero to have a multiplicative inverse.

## Axiom 8.3

For every integer $x \neq 0$ there is a unique number represented by $\bar{x}$ called the *reciprocal of x* and a binary operation $\otimes$ called *multiplication* for which $x \otimes \bar{x} = 1$.

*Extending Number Systems* 141

Although we are accustomed to writing the reciprocals of

$$2, 3, 4, 5, \cdots$$

as

$$\frac{1}{2}, \frac{1}{3}, \frac{1}{4}, \frac{1}{5}, \cdots$$

the numeral 1 is distracting and not at all essential to our purpose, so we shall use the simpler notation

$$\{\cdots \bar{3}, \bar{2}, \bar{2}, \bar{1}, \bar{2}, \bar{3}, \bar{4}, \cdots\}$$

to represent the reciprocals of the non-zero integers. Since

$$1 \otimes \bar{1} = 1 \quad \text{and} \quad 1 \times 1 = 1$$

it seems possible to identify $\bar{1}$ as 1 and let

$$x \otimes y = x \cdot y \quad \text{for} \quad x, y \in J$$

Also, if the new numbers are to be combined with the integers, we need to discover the nature of the resulting numbers.

## Axiom 8.4

There is a binary operation $\oplus$ called *addition* on the combined set of integers and reciprocals. For all $m, n \in J, m \oplus n = m + n$ and $m \otimes n = m \times n$. $\oplus$ and $\otimes$ are associative and commutative, and $\otimes$ is distributive over $\oplus$.

With these familiar properties, such observations as

$$1 = 1 \cdot 1 = (x \otimes \bar{x}) \otimes (y \otimes \bar{y}) = x \otimes (\bar{x} \otimes y) \otimes \bar{y}$$
$$= x \otimes \bar{y} \otimes (y \otimes \bar{x}) = (x \otimes \bar{y}) \otimes (y \otimes \bar{x})$$

show the need to consider combinations like

$$m \otimes \bar{n}$$

This combination becomes especially significant when we observe that every integer can be expressed in the form

$$m = m \times 1 = m \otimes 1 = m \otimes \bar{1}$$

and that

$$\bar{n} = 1 \otimes \bar{n}$$

for every reciprocal $\bar{n}$ since

$$1 = n \otimes \bar{n} = (n \times 1) \otimes \bar{n} = n \otimes (1 \otimes \bar{n})$$

which identifies $1 \otimes \bar{n}$ as $\bar{n}$.

*Definition 8.4.* The elements of the set $F = \{m \otimes \bar{n} \mid m, n \in J, n \neq 0\}$ are called *rational numbers*.

The observation that
$$(m \otimes \bar{n}) \otimes n = m \otimes (\bar{n} \otimes n) = m \otimes 1 = m$$
brings to memory a similar statement,
$$(m \div n) \times n = m$$
showing the relation
$$m \div n = m \otimes \bar{n}$$
which holds when both expressions are defined. We shall have more to say about the more familiar interpretations of rational numbers after we examine briefly the properties of the system $(F; \oplus, \otimes)$.

In considering the product of two elements of $F$,
$$(a \otimes \bar{b}) \otimes (c \otimes \bar{d}) = (a \cdot c) \otimes (\bar{b} \otimes \bar{d})$$
we see that the numbers can be arranged so that the integers are multiplied [an operation in $(J; +, \times)$] and the two reciprocals are multiplied. Since $J$ is closed under $\times$, $a \cdot c$ is an integer. If $\bar{b} \otimes \bar{d}$ were a reciprocal then $(a \otimes \bar{b}) \otimes (c \otimes \bar{d})$ would be a rational number and $F$ would be closed under $\otimes$. Now

$$\bar{b} \otimes \bar{d} = \bar{b} \otimes \bar{d} \otimes 1 = \bar{b} \otimes \bar{d} \otimes [(bd) \otimes \overline{bd}]$$
$$= \bar{b} \otimes \bar{d} \otimes b \otimes d \otimes \overline{bd}$$
$$= (b \otimes \bar{b}) \otimes (d \otimes \bar{d}) \otimes \overline{bd}$$
$$= 1 \otimes 1 \otimes \overline{bd}$$
$$= \overline{bd}$$

which shows that $b \cdot d$ is a non-zero integer (since $b \neq 0$ and $d \neq 0$) so that $\bar{b} \otimes \bar{d}$ is indeed a reciprocal. This investigation yields three true statements which, in our treatment, illustrate the three basic types of mathematical statements subject to proof. A *lemma* is a statement proved true so that it can be used to prove a theorem. A *theorem* is a provable statement which is of broader significance in that it is useful on a number of occasions. A *corollary* is a statement implied directly or almost directly by the theorem which it follows. Compare these descriptions with the following sequence:

## Lemma

$\bar{m} \otimes \bar{n} = \overline{m \cdot n}$ for all non-zero $m, n \in J$.

## Theorem 8.6

$(a \otimes \bar{b}) \otimes (c \otimes \bar{d}) = (a \cdot c) \otimes \overline{b \cdot d}$ for $a \otimes \bar{b}, c \otimes \bar{d} \in F$.

*Corollary.* $F$ is closed under $\otimes$.

## Extending Number Systems

We will furnish a few examples so you can verify your grasp of the ideas presented thus far.

EXAMPLE: According to the lemma
$$\overline{{}^-2} \otimes \overline{3} = \overline{{}^-2 \cdot 3} = \overline{{}^-6}$$
and
$$\overline{5} \otimes \overline{6} = \overline{5 \cdot 6} = \overline{30}$$
where each product of reciprocals is identified by the knowledge of the result of $^-2 \cdot 3$ and $5 \cdot 6$ in the system $(J; +, \times)$ of integers.

EXAMPLE: The product
$$(^-2 \otimes \overline{3}) \otimes (5 \otimes \overline{6}) = {}^-2 \otimes \overline{3} \otimes 5 \otimes \overline{6} = (^-2 \otimes 5) \otimes (\overline{3} \otimes \overline{6})$$
$$= (^-2 \cdot 5) \otimes \overline{3 \cdot 6} = {}^-10 \otimes \overline{18}$$

was obtained and identified as a rational number by using the properties of $\otimes$ and the lemma. Time and effort are saved if Theorem 8.6 is applied directly to give
$$(^-2 \otimes \overline{3}) \otimes (5 \otimes \overline{6}) = (^-2 \cdot 5) \otimes \overline{3 \cdot 6} = {}^-10 \otimes \overline{18}$$

We turn our attention now to the problem of finding the sum of two rational numbers, hoping to find a method which will again allow the use of our knowledge of the familiar system $(J; +, \times)$. The form

$$a \otimes \overline{b} + c \otimes \overline{d}$$

reminds us of the distributive property of $\otimes$ over $\oplus$. To make use of this property, however, we need a repeated multiplier or common factor, as it is called. The traditional method for finding the common factor is to change the form of the two rational numbers so that their numerals have the same reciprocal factors, whence the principle of distributivity may be applied as follows:

$$a \otimes \overline{b} \oplus c \otimes \overline{d} = a \otimes \overline{b} \otimes 1 \oplus c \otimes \overline{d} \otimes 1$$
$$= a \otimes \overline{b} \otimes (d \otimes \overline{d}) \oplus c \otimes \overline{d} \otimes (b \otimes \overline{b})$$
$$= (a \otimes d) \otimes (\overline{b} \otimes \overline{d}) \oplus (b \otimes c) \otimes (\overline{b} \otimes \overline{d})$$
$$= (a \cdot d) \otimes \overline{b \cdot d} \oplus (b \cdot c) \otimes \overline{b \cdot d}$$
$$= (ad + bc) \otimes \overline{bd}$$

Since $J$ is closed under $\times$ and $+$, $ad + bc$ and $bd$ are integers with $bd \neq 0$, so that we can perform addition in $(F; +, \times)$ by performing operations in $(J; +, \times)$.

## Theorem 8.7

$$a \otimes \bar{b} \oplus c \otimes \bar{d} = (ad + bc) \otimes \overline{bd} \text{ for all } a \otimes \bar{b}, c \otimes \bar{d} \in F.$$

*Corollary.*   $F$ is closed under $\oplus$.

EXAMPLE: Using the method demonstrated above,

$$\begin{aligned}
{}^-2 \otimes \bar{3} \oplus 5 \otimes \bar{6} &= {}^-2 \otimes \bar{3} \otimes (6 \otimes \bar{6}) \oplus 5 \otimes \bar{6} \otimes (3 \otimes \bar{3}) \\
&= ({}^-2 \cdot 6) \otimes \overline{3 \cdot 6} \oplus (5 \cdot 3) \otimes \overline{6 \cdot 3} \\
&= ({}^-12 + 15) \otimes \overline{18} \\
&= 3 \otimes \overline{18}
\end{aligned}$$

If you recognize the result of the example as the fraction

$$\frac{3}{18}$$

you probably wish to see its final form as

$$\frac{1}{6} \quad \text{or} \quad 1 \otimes \bar{6}$$

since we are accustomed to leaving fractions in reduced form, and because

$$\frac{3}{18} = \frac{1}{6}$$

This raises the final question which we will consider in this section: When does

$$a \otimes \bar{b} = c \otimes \bar{d}$$

Let us first suppose that there are two such different symbols which represent the same rational number and try to deduce a corresponding relation involving the four integers. If

$$a \otimes \bar{b} = c \otimes \bar{d}$$

then

$$a \otimes \bar{b} \otimes (b \otimes d) = c \otimes \bar{d} \otimes (b \otimes d)$$

that is,

$$(a \cdot d) \otimes (b \otimes \bar{b}) = (b \cdot c) \otimes (d \otimes \bar{d})$$

or

$$(a \cdot d) \cdot 1 = (b \cdot c) \cdot 1$$

whence

$$a \cdot d = b \cdot c$$

which can be ascertained entirely within $(J; +, \times)$. By starting with

$$a \cdot d = b \cdot c$$

*Extending Number Systems* 145

and multiplying by $\bar{b} \otimes \bar{d}$ both $a \cdot d$ and $b \cdot c$, it can be shown that

$$a \otimes \bar{b} = c \otimes \bar{d}$$

but this will be left as an exercise. These results are summarized by the following theorem, which is often used as a definition.

## Theorem 8.8

For integers $a, b, c, d$ with $b \neq 0 \neq d$, $a \otimes \bar{b} = c \otimes \bar{d}$ if and only if $a \cdot d = b \cdot c$.

EXAMPLE: $5 \otimes \bar{3} = 15 \otimes \bar{9}$

because

$$5 \cdot 9 = 3 \cdot 15.$$

Similarly, since

$$3 \cdot 6 = 18 \cdot 1$$

it is also true that

$$3 \otimes \bar{18} = 1 \otimes \bar{6}$$

## Exercise 8.3

Let $J_7 = \{0, 1, 2, 3, 4, 5, 6\}$ and let $+$ and $\times$ represent *modulo* 7 operations (like clock addition and multiplication on a seven hour clock face starting at 0). For example, $3 + 4 = 0$, $5 + 6 = 4$, and $3 \times 4 = 5$. Use the tables for $+$ and $\times$ to answer questions 1–5, and note that 0 and 1 are the identity elements for $+$ and $\times$ respectively.

| + | 0 | 1 | 2 | 3 | 4 | 5 | 6 |
|---|---|---|---|---|---|---|---|
| 0 | 0 | 1 | 2 | 3 | 4 | 5 | 6 |
| 1 | 1 | 2 | 3 | 4 | 5 | 6 | 0 |
| 2 | 2 | 3 | 4 | 5 | 6 | 0 | 1 |
| 3 | 3 | 4 | 5 | 6 | 0 | 1 | 2 |
| 4 | 4 | 5 | 6 | 0 | 1 | 2 | 3 |
| 5 | 5 | 6 | 0 | 1 | 2 | 3 | 4 |
| 6 | 6 | 0 | 1 | 2 | 3 | 4 | 5 |

| × | 0 | 1 | 2 | 3 | 4 | 5 | 6 |
|---|---|---|---|---|---|---|---|
| 0 | 0 | 0 | 0 | 0 | 0 | 0 | 0 |
| 1 | 0 | 1 | 2 | 3 | 4 | 5 | 6 |
| 2 | 0 | 2 | 4 | 6 | 1 | 3 | 5 |
| 3 | 0 | 3 | 6 | 2 | 5 | 1 | 4 |
| 4 | 0 | 4 | 1 | 5 | 2 | 6 | 3 |
| 5 | 0 | 5 | 3 | 1 | 6 | 4 | 2 |
| 6 | 0 | 6 | 5 | 4 | 3 | 2 | 1 |

1. If $^-n$ for $n \in J_7$ means that $n + {^-n} = 0$, find
   (a) $^-1$     (c) $^-3$     (e) $^-5$
   (b) $^-2$     (d) $^-4$     (f) $^-6$

**2.** If $\bar{n}$ for $n \in J_7$ means that $n \times \bar{n} = 1$, find
(a) $\bar{1}$   (c) $\bar{3}$   (e) $\bar{5}$
(b) $\bar{2}$   (d) $\bar{4}$   (f) $\bar{6}$

**3.** From the results of Problems 1 and 2 and using the tables to find sums and products, calculate the following as elements of $J_7$:
(a) $3 \times \bar{4}$   (b) $(2 \times \bar{6}) \times (5 \times \bar{4})$   (c) $3 \times \bar{5} + 4 \times \bar{2}$

**4.** Use Theorems 8.6 and 8.7 and the tables to calculate
(a) $(2 \times \bar{6}) \times (5 \times \bar{4})$   (b) $3 \times \bar{5} + 4 \times \bar{2}$
Check your results against those of 3(b) and 3(c).

**5.** Show that $5 \times \bar{4} = 4 \times \bar{6}$ by using
(a) results of Problem 2 and the $\times$ table,
(b) Theorem 8.8 and the $\times$ table.

**6.** Use the theory developed in this section to make the following calculations in $(F; +, \times)$.
(a) $\bar{5} \times \bar{8}$
(b) $(3 \times \bar{6}) \times (5 \times \bar{2})$
(c) $(1 \times \bar{1}) \times (a \times \bar{b})$
(d) $3 \times \overline{11} + 4 \times \overline{11}$
(e) $0 \times \bar{5} + 4 \times \bar{3}$
(f) $^-5 \times \bar{3} + 7 \times \overline{^-2}$

**\*7.** Prove: If $c$ and $d$ are non-zero integers then $(c \times \bar{d}) \times (d \times \bar{c}) = 1$.

**\*8.** Prove: If $b \times \bar{c} \in F$ and $d \neq 0$ is an integer, then $(b \cdot d) \times \overline{c \cdot d} = b \times \bar{c}$.

**\*9.** (a) Show that $0 \times \bar{b} = 0 \times \bar{c}$ for all non-zero integers $b$ and $c$.
(b) Use $0 = 0 \cdot 1 = 0 \times \bar{1}$ to prove that zero is the identity for $\oplus$ in $(F; \oplus, \otimes)$.

**\*10.** Prove that if integers $a, b, c, d$ exist such that $a \cdot d = b \cdot c$ and $b \neq 0 \neq d$, then $a \otimes \bar{b} = c \otimes \bar{d}$.

**\*11.** Show that for $m, n \in J$ (a) $m \times \bar{1} + n \times \bar{1} = m + n \in J$ and (b) $(m \times \bar{1}) \times (n \times \bar{1}) = m \cdot n \in J$, thus establishing the facts essential to the proposition that the system $(J; +, \times)$ is imbedded in the system $(F; +, \times)$.

## RATIONAL NUMBERS AND COMMON FRACTIONS

Now that we have indicated the nature of a mathematical development of the system of rational numbers as an extension of the system of integers, we turn to the more traditional interpretations and algorithms. The basic form of numeration habitually used is called a *common fraction*

$$\frac{b}{c}$$

which is related to the form of the previous section by the equality

$$b \otimes \bar{c} = \frac{b}{c}$$

The following definition gives some of the symbols and corresponding words commonly used to represent rational numbers. However, most of these symbols and terms refer more explicitly to the nature of the symbol than to the types of numbers which they represent.

*Definition 8.5.* The equivalent symbols $a/b = a \div b = a : b$ representing $a \times \bar{b}$ are called respectively *common fraction*, *quotient*, and *ratio* (of $a$ to $b$). $a$ is the *numerator* and $b$ the *denominator* of the fraction $a/b$, while in the quotient $a \div b$ they are called *dividend* and *divisor* respectively.

# WORD ORIGINS

From time to time we have emphasized the self-explanatory nature of well-chosen terminology and the importance of developing the habit of analyzing the common meanings of words in order to gain insight into their specialized usage as technical terms. Certain parts of Definition 8.5 afford excellent examples of this principle. Perhaps you have already recognized that the expression *rational number* comes from the *ratio* concept.

The word *fraction* comes from a Latin word meaning *to break* (compare with *fracture*) and reflects the idea that a number like 1/3, for example, represents only a part of the unit so that a physical object such as a yardstick would have to be *broken* in order to have a foot-long stick to represent one third. Of course this is a more primitive concept of the positive rational numbers than we have used, but it has been the sole point of view used to introduce rational numbers in the past and may still be one of the best intuitive devices for introductory work with children. In this context the idea of $a/b$ as a fraction is not very meaningful unless we restrict $a$ and $b$ to the natural numbers.

You may recall in our intuitive investigation of division that we made a physical model by partitioning a set into equivalent subsets as shown for $15 \div 5$, illustrated below. From this viewpoint, we would associate the number three with each of the five subsets shown. However, if the number *one* had

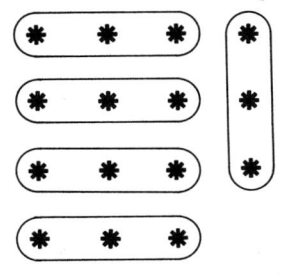

been associated with the original set instead of the number fifteen, then we would have associated with each subset the number *one fifth* and the symbol

$1/5 (= 1 \div 5)$. A subset consisting of two of the equivalent subsets would be associated with the number *two fifths*, symbolized by

$$\frac{2}{5}$$

In an effort to instill a sense of reality into these abstract ideas, pictures of objects partitioned into congruent (look-alike) sections have been presented. Although "pies" (circles with interiors) offer excellent examples, they are more difficult to partition than line segments, pointing out another advantage of the number line. The pictures below show some typical pictorial representations of 2/5 which vary somewhat in the degree of abstraction. From this viewpoint, the 2 is emphasized as the *number* of parts being represented and the *fifth*

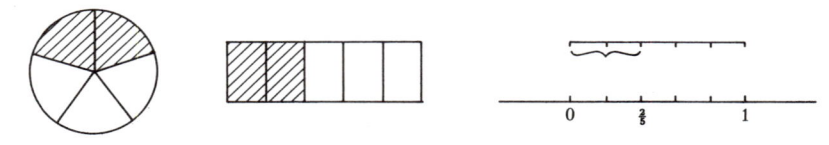

names the *kind* of parts. Thus 2 is the *numerator* (or *numberer*) and *five* is the denominator (or *namer*). This somewhat homely description can be quite helpful in recalling the details of the addition algorithm soon to be discussed.

## THE ALGORITHMS IN $(F; +, \times)$

In the preceding section we discovered ways of performing the basic operations of $(F; +, \times)$ by means of those of $(J; +, \times)$ and described them in Theorems 8.6 and 8.7. We also found that each element of $F$ can be represented in many ways (Problem 8), so we shall first search for a standard symbol for each element of $F$ and see how this idea modifies the algorithms for multiplication and addition. Since the principle stated in Problem 8 of Exercise 8.3 is quite important to the theory and practice of $(F; +, \times)$ we will state it as a theorem and offer a proof.

## *Theorem 8.9*

If $(a/b) \in F$, $c \in J$ and $c \neq 0$, then $\dfrac{ac}{bc} = \dfrac{a}{b}$.

*Proof:* If $(a/b) \in F$ then $a/b = a \times \bar{b}$. If $c \in J$ with $c \neq 0$ then $c \times \bar{c} = 1$. Then by the properties of $(F; +, \times)$, $a/b = a \times \bar{b} = a \times \bar{b} \times 1 = a \times \bar{b} \times (c \times \bar{c}) = (a \times c) \times (\bar{b} \times \bar{c}) = (a \cdot c) \times \overline{b \cdot c} = ac/bc$.

EXAMPLES: It should be observed that, like the distributive property, this property of Theorem 8.9 allows equivalent fractions with smaller terms (numerator and denominator) as well as those whose terms are larger. For example, not only is

$$\frac{9}{15} = \frac{9 \cdot 2}{15 \cdot 2} = \frac{18}{30}$$

but also

$$\frac{9}{15} = \frac{3 \cdot 3}{5 \cdot 3} = \frac{3}{5}$$

It should also be clear from the nature of $(J; +, \times)$ and from Theorem 8.9 that every rational number can be represented by a fraction $a/b$ where $b$ is a positive integer. For example, if the number can be represented by

$$\frac{15}{-19}$$

then it can also be represented by

$$\frac{15 \cdot (-1)}{-19 \cdot (-1)} = \frac{-15}{19}$$

according to Theorem 8.9.

*Definition 8.6.* The fraction $a/b$ is a *standard* fraction iff $a \in J, b \in N$ and 1 is the only natural number which divides both $a$ and $b$.

## NUMBER THEORY

A practical question which arises from this discussion is, "How can it be decided whether there is a natural number other than 1 which will divide both terms of the fraction?" For fractions like

$$\frac{9}{15} \quad \text{or} \quad \frac{3}{5}$$

the question can be resolved rather quickly by inspection, but for a fraction such as

$$\frac{-391}{437}$$

this is not an idle question. The branch of mathematics which deals with basic questions of this type and is limited to investigations of the properties of $J$ or $W$ is called *Number Theory*. The division algorithm is one of the basic theorems of number theory, and we will mention two others which are directly related to the question. One of these is the most direct answer to the question and is a direct consequence of the division algorithm. We will illustrate it using the numbers 391 and 437 of the fraction mentioned earlier. The negative sign can be handled separately later, as you will see.

The relations shown in the computations

$$391 \overline{)437}^{\ 1} = 391 \cdot 1 + 46 \quad \text{or} \quad 46 = 437 - 391$$
$$\phantom{391 \overline{)437}} \, 391$$
$$\phantom{391 \overline{)437}} \overline{\phantom{391}}$$
$$\phantom{391 \overline{)437}} \ \ 46$$

show that any number which divides both 391 and 437 will also divide 46 = 437 − 391 (distributive property) and any number which divides both 46 and 391 will also divide 437 = 391 + 46. Therefore, we transfer our attention to 46 and 391, whence

$$46 \overline{)391}^{\ 8} = 46 \cdot 8 + 23 \quad \text{or} \quad 23 = 391 - 46 \cdot 8$$
$$\phantom{46 \overline{)391}} \, 368$$
$$\phantom{46 \overline{)391}} \overline{\phantom{368}}$$
$$\phantom{46 \overline{)391}} \ \ 23$$

which shows that any number which divides both 46 and 391 will also divide 23 and any number which will divide both 23 and 46 will also divide 391 (and subsequently 437 also, by the previous argument). Continuing,

$$23 \overline{)46}^{\ 2} = 23 \cdot 2 + 0$$
$$\phantom{23 \overline{)46}} \, 46$$
$$\phantom{23 \overline{)46}} \overline{\phantom{46}}$$
$$\phantom{23 \overline{)46}} \ \ \, 0$$

so that a remainder zero is eventually obtained since each remainder is used as the next divisor and the next remainder must be less than the new divisor. From the argument, the last non-zero remainder, 23, divides each of the original numbers 391 and 437,

$$23 \overline{)391}^{\ 17} \qquad 23 \overline{)437}^{\ 19}$$
$$\phantom{23 \overline{)391}} 230 \phantom{xxxx} 230$$
$$\phantom{23 \overline{)391}} \overline{\phantom{230}} \phantom{xxxx} \overline{\phantom{230}}$$
$$\phantom{23 \overline{)391}} 161 \phantom{xxxx} 207$$
$$\phantom{23 \overline{)391}} 161 \phantom{xxxx} 207$$
$$\phantom{23 \overline{)391}} \overline{\phantom{161}} \phantom{xxxx} \overline{\phantom{207}}$$

so that

$$\frac{-391}{437} = \frac{-17 \cdot 23}{19 \cdot 23} = \frac{-17}{19}$$

In the standard development of this theory it is proved that this method produces the *largest* number which will divide both numbers. In the realm of algebra the following definition would be inadequate, but for our purposes, it is more meaningful than the more general definition.

*Definition 8.7.* The *greatest common divisor*, also called *highest common factor*, of a set of integers is the largest natural number which divides all integers of the set. If that number is 1 then the integers are said to be *relatively prime*.

In the following exercise we will use Theorem 8.9 extensively. In order to apply the theorem to realistic situations, we will first state another theorem which simplifies multiplication and addition in $(F; +, \times)$ using common fractions. The proof of the theorem is also rather simple when elements are written in the $\overline{b \otimes c}$ form, so it is left as an exercise.

## *Theorem 8.10*

If $a/b$, $c/b$ and $c/d$ are elements of $F$ then

$$(1) \quad \frac{a}{b} + \frac{c}{b} = \frac{a+c}{b} \quad \text{and} \quad (2) \quad \frac{a}{b} \cdot \frac{c}{d} = \frac{a \cdot c}{b \cdot d}$$

Notice how part (1) of the theorem follows the principle of place value. For example,

$$\frac{3}{7} + \frac{2}{7} = \frac{3+2}{7} = \frac{5}{7}$$

or 3 sevenths + 2 sevenths = (3 + 2) sevenths = 5 sevenths just as in the case of digits 3 and 2 in 3 hundreds + 2 hundreds = (3 + 2) hundreds. The distributivity of $\times$ over $+$ is the common property here. The combination of part (2) with Theorem 8.9 produces an interesting observation. For example, the product

$$\frac{5}{6} \times \frac{9}{10} = \frac{5 \cdot 9}{6 \cdot 10}$$

can be written as

$$\frac{3 \cdot 3 \cdot 5}{2 \cdot 2 \cdot 3 \cdot 5} = \frac{3 \cdot 15}{4 \cdot 15} = \frac{3}{4}$$

the last of which is a standard fraction.

## Exercise 8.4

*1. Write a common fraction to represent each of the following:
(a) $^-5 \times 8$  (b) $3 \times {}^-6$  (c) $^-6 \times {}^-2$  (d) $^-4 \div 6$
(e) $9 : 6$

*2. Shade a portion of the rectangular region to picture the fraction
(a) $\frac{3}{8}$   (b) $\frac{4}{6}$
(c) $\frac{1}{3}$   (d) $\frac{1}{2}$

*3. Write a common fraction for each of the points $a, b, c, d, e$ on the number line below.

*4. Write the standard fractions which represent
(a) $^-6 : 8$  (b) $\frac{15}{-24}$  (c) $\frac{1}{2} \div 3$  (d) $\frac{-5}{5}$  (e) $\frac{0}{2}$

*5. Fill in the missing numerals to make true statements, if possible.
(a) $\frac{6}{15} = \frac{2}{\phantom{x}}$  (b) $\frac{8}{12} = \frac{\phantom{x}}{-6}$  (c) $\frac{0}{-5} = \frac{\phantom{x}}{2}$  (d) $^-3:7 = 9 \div$
(e) $\frac{4}{0} = {}^-2 :$

6. Demonstrate on number lines
(a) $\frac{2}{3} + \frac{2}{3}$  (b) $\frac{3}{10} + \frac{4}{10}$  (c) $\frac{5}{6} - \frac{7}{6}$  (d) $\frac{-3}{10} + \frac{-2}{10}$

7. Find the greatest common divisors of the following sets.
(a) $\{112, 126\}$  (b) $\{^-43, 37\}$  (c) $\{^-108, 117, 126\}$

Write all answers as standard fractions in the following problems.

*8. (a) $\frac{112}{126} =$  (b) $\frac{-108}{117} =$  (c) $\frac{-108}{-126}$

*9. Use Theorem 8.10 to calculate:
(a) $\frac{4}{13} + \frac{-9}{13}$  (b) $\frac{-5}{28} + \frac{9}{-28}$  (c) $\frac{2}{5} + \frac{4}{15}$  (d) $\frac{2}{5} + \frac{1}{3} + \frac{1}{-15}$

*10. Calculate by Theorem 8.10:
(a) $\frac{2}{7} \cdot \frac{3}{7}$  (b) $\frac{-10}{21} \cdot \frac{14}{15}$  (c) $\frac{0}{-3} \cdot \frac{-6}{-5}$

## PRIME NUMBERS

Another concept of number theory, even more basic in its application, is the idea of *prime number*. Something of its nature and utility is apparent by the comparison of the two products

$$21 \cdot 22 \quad \text{and} \quad 14 \cdot 33$$

At first glance, there should be no reason to suspect that both numerals represent the same number. Although this fact may be verified by performing both multiplications, there is a much more informative and useful method. The observation that

$$21 \cdot 22 = 3 \cdot 7 \cdot 2 \cdot 11$$

and

$$14 \cdot 33 = 2 \cdot 7 \cdot 3 \cdot 11$$

should be just as convincing, since the same basic set of numbers is to be multiplied in each case. Furthermore, if the numbers had not been the same, this method would not only reveal it but, in a sense, would show the reason. Suppose, for example, that we have to find

$$\frac{11}{7 \cdot 16} + \frac{-13}{21 \cdot 6}$$

If we write $2 \cdot 2 \cdot 2 \cdot 2$ for 16, $3 \cdot 7$ for 21 and $2 \cdot 3$ for 6, we have

$$\frac{11}{7 \cdot 2 \cdot 2 \cdot 2 \cdot 2} + \frac{-13}{3 \cdot 7 \cdot 2 \cdot 3}$$

whence the $7 \cdot 16$ and $21 \cdot 6$ are clearly not the same. However, by applying Theorem 8.9 we have

$$\frac{11 \cdot 3 \cdot 3}{7 \cdot 2 \cdot 2 \cdot 2 \cdot 2 \cdot 3 \cdot 3} + \frac{-13 \cdot 2 \cdot 2 \cdot 2}{3 \cdot 7 \cdot 2 \cdot 3 \cdot 2 \cdot 2 \cdot 2}$$

in which the denominators are the same. We pause briefly to examine the nature of these numbers before continuing with their application.

*Definition 8.8.* A whole number is a *composite* number if it is the product of two smaller whole numbers. A whole number is a *prime* number if it is greater than 1 and is not a composite number.

It follows from Definition 8.8 that every element of $W$ is either an identity element, a prime, or a composite number exclusively. Since every negative integer is the product of $^-1$ and a whole number, it is not necessary for our purposes to define prime integer, although the concept of a prime element of a set has been defined for systems far more abstract and sophisticated than

$(J; +, \times)$. You should recognize the set

$$P = \{2, 3, 5, 7, 11, 13, 17, 19, 23, 29, 31, \cdots\}$$

as the set of all prime whole numbers, and you should be able to select as many of them as you choose without overlooking any. The usefulness of prime numbers lies in the following statement, which is often called *the fundamental theorem of arithmetic.*

## Theorem 8.11

Every whole number greater than one is the product of a unique set of primes.

To state this theorem simply, we have taken two liberties: (1) a single prime is considered to be a product, and (2) a set may contain an element more than once. For example, the set of factors of 17 is

$$\{17\}$$

while that for 42 is

$$\{2, 3, 7\}$$

and the unique set of prime factors of 36 is

$$\{2, 2, 3, 3\}$$

From Theorem 8.10 it should be clear that if $p$ is a *prime factor* of $r$ (that is, $r = p \cdot q$) then any other prime factor of $r$ must be a factor of $q$. Although this principle is also basic to the study of number theory, we will not state the theorem but simply demonstrate its usefulness in finding the prime factors of a number.

EXAMPLE: Write 2,394 as a product of primes. We first try to divide 2,394 by the least prime, 2, and obtain 1,197 so that any further prime factors must be divisors of 1,197. Since 2 is not a factor of 1,197 we try the next larger prime, 3, obtaining 399 which can also be divided by 3, giving 133 as a quotient. Since 133 is not divisible by 3, we try 5 (the next larger prime) with no success. Then we try 7 and find the other factor to be 19 which we recognize as a prime number, so the procedure is finished. By successive division we have found the prime factors of 2,394 to be 2, 3, 3, 7 and 19 so that

$$2{,}394 = 2 \cdot 3 \cdot 3 \cdot 7 \cdot 19$$

Although a great amount of literature deals with prime numbers, our major concern with them will be their usefulness for finding the standard fraction of a rational number and in finding the least common denominator to add rational numbers. We shall give one more example for each operation in which the factoring is not trivial.

EXAMPLE:

$$\frac{182}{247} \cdot \frac{57}{49} = \frac{(2 \cdot 7 \cdot 13)(3 \cdot 19)}{(13 \cdot 19)(7 \cdot 7)}$$

$$= \frac{(2 \cdot 3)(7 \cdot 13 \cdot 19)}{7(7 \cdot 13 \cdot 19)}$$

$$= \frac{2 \cdot 3}{7}$$

$$= \frac{6}{7}$$

EXAMPLE:

$$\frac{27}{182} + \frac{^-45}{247} = \frac{27}{2 \cdot 7 \cdot 13} + \frac{^-45}{13 \cdot 19}$$

$$= \frac{27 \cdot 19}{2 \cdot 7 \cdot 13 \cdot (19)} + \frac{^-45 \cdot (2 \cdot 7)}{13 \cdot 19 \cdot (2 \cdot 7)}$$

$$= \frac{513}{2 \cdot 7 \cdot 13 \cdot 19} + \frac{^-630}{2 \cdot 7 \cdot 13 \cdot 19}$$

$$= \frac{^-117}{2 \cdot 7 \cdot 13 \cdot 19}$$

$$= \frac{^-9 \cdot (13)}{2 \cdot 7 \cdot 19 \cdot (13)}$$

$$= \frac{^-9}{266}$$

## INVERSE OPERATIONS IN $(F; +, \times)$

As we continue to investigate and compare the properties of the rational number system with those of the integers, a question arises concerning the existence of operations inverse to $\oplus$ and $\otimes$. In addition, since it has already been established that every pair of rational numbers can be represented by fractions having the same denominator, we need consider only such fractions.

If
$$\frac{a}{c} \ominus \frac{b}{c}$$
is to have meaning by Definition 7.3, where $\ominus$ is inverse to $\oplus$, then
$$\left(\frac{a}{c} \ominus \frac{b}{c}\right) \oplus \frac{b}{c} = \frac{a}{c}$$
But by the same definition and the properties of $(J; +, \times)$
$$\frac{a}{c} = \frac{(a-b)+b}{c}$$
and by Theorem 8.10
$$\frac{(a-b)+b}{c} = \frac{a-b}{c} + \frac{b}{c}$$
From these relations we see that
$$\left(\frac{a}{c} \ominus \frac{b}{c}\right) \oplus \frac{b}{c} = \frac{a-b}{c} \oplus \frac{b}{c}$$
which implies
$$\frac{a}{c} \ominus \frac{b}{c} = \frac{a-b}{c}$$
Since $a - b = a + {}^-b$ is an integer for all $a, b \in J$ it follows that subtraction is defined in $(F; +, \times)$ and $F$ is closed under subtraction.

## Theorem 8.12

The operation $\ominus$ on $F$ called *subtraction* and defined by
$$\frac{a}{c} \ominus \frac{b}{c} = \frac{a-b}{c}$$
is inverse to addition in $(F; +, \times)$.

*Corollary.*   $F$ is closed under subtraction.

EXAMPLE: $\quad \dfrac{7}{13} \ominus \dfrac{{}^-4}{13} = \dfrac{7 - {}^-4}{13} = \dfrac{7+4}{13} = \dfrac{11}{13}$

Changing to the conventional subtraction symbol for a second less trivial problem, we have

$$\frac{5}{-18} - \frac{7}{12} = \frac{-5}{2 \cdot 3 \cdot 3} - \frac{7}{2 \cdot 2 \cdot 3}$$

$$= \frac{-5 \cdot (2)}{2 \cdot 3 \cdot 3 \cdot (2)} - \frac{7 \cdot (3)}{2 \cdot 2 \cdot 3 \cdot (3)}$$

$$= \frac{-10 - 21}{2 \cdot 2 \cdot 3 \cdot 3}$$

$$= \frac{-10 + -21}{2 \cdot 2 \cdot 3 \cdot 3}$$

$$= \frac{-31}{36}$$

Since the existence of an inverse operation has led us in the past to the analogous concept of inverse elements, we now check to see if there is an element $x \in F$ for which

$$\frac{b}{c} + x = 0$$

Since we know that

$$0 = \frac{0}{c} = \frac{b + -b}{c} = \frac{b}{c} + \frac{-b}{c}$$

we conclude that $-b/c$ is the *negative* of $b/c$.

## Theorem 8.13

For each $(b/c) \in F$ for which $b/c \neq 0$, there is a unique additive inverse element symbolized by $-(b/c)$ and identified by $-(b/c) = -b/c$.

Combining these results gives a familiar relation appearing in a more generalized situation:

$$\frac{a}{c} \ominus \frac{b}{c} = \frac{a-b}{c} = \frac{a+-b}{c} = \frac{a}{c} \oplus \frac{-b}{c} = \frac{a}{c} \oplus {-}\left(\frac{b}{c}\right)$$

or, using the more conventional symbols for subtraction and addition,

$$\frac{a}{c} - \frac{b}{c} = \frac{a}{c} + {-}\left(\frac{b}{c}\right)$$

In seeking to establish a case for division in $(F; +, \times)$ we must have

$$\left(\frac{a}{b} \oplus \frac{c}{d}\right) \otimes \frac{c}{d} = \frac{a}{b}$$

If $a/b \neq 0$ then certainly $c/d \neq 0$ (or $c \neq 0$) provided $a/b \div c/d$ is defined. Now

$$\frac{a}{b} = \frac{a}{b} \otimes 1 = \frac{a}{b} \otimes \left(\frac{d}{c} \otimes \frac{c}{d}\right)$$

using a result of Problem 7 in Exercise 8.3, so that

$$\frac{a}{b} = \left(\frac{a}{b} \otimes \frac{d}{c}\right) \otimes \frac{c}{d} = \left(\frac{a}{b} \oplus \frac{c}{d}\right) \otimes \frac{c}{d}$$

which identifies

$$\frac{a}{b} \oplus \frac{c}{d} \quad \text{as} \quad \frac{a}{b} \otimes \frac{d}{c}$$

an element of $F$.

### Theorem 8.14

If $\dfrac{a}{b}, \dfrac{c}{d} \in F$ with $c \neq 0$, then $\dfrac{a}{b} \oplus \dfrac{c}{d} = \dfrac{a}{b} \times \dfrac{d}{c}$

*Corollary.* $F$ is closed under division by non-zero elements.

## ORDER IN $(J; +, \times)$ AND $(F; +, \times)$

Our attention in the development of the systems of integers and rational numbers has been focused on the nature of the numbers relative to the operations. Although we have discussed *equality* of fractions, we have not yet examined the way in which two different numbers may be related, as pictured, for example on the number line. In Chapter 2, while we were still thinking of the elements of $W$ as cardinal numbers of sets, we gave a definition for a *greater than* relation. The intuitive concept which makes us agree that

$$13 > 6$$

is that "something has to be added to 6 to get 13;" that is,

$$13 > 6 \quad \text{because} \quad 13 = 6 + 7$$

However, it is true that

$$13 = 15 + {}^-2$$

*Extending Number Systems*

so that 13 would be greater than 15 also by this argument. We can thus see that it is necessary to specify that a *positive* number can be added to the lesser number to produce the greater.

All of this leads to the question, "What is a positive number?" We have already defined the natural numbers $\{1, 2, 3, \cdots\}$ as *positive integers*. Which rational numbers should be considered positive? If possible, we should use a criterion which will also apply to any "new" numbers which we may later invent. In searching for a characteristic which distinguishes the set of positive integers from the set of negative integers, we note that $N$ is closed under both $+$ and $\times$ while $\{^-1, ^-2, ^-3, \cdots\}$ is closed under $+$ but *not* under $\times$. We are encouraged to make the following definition of a property which is often refered to as the *trichotomy* property.

*Definition 8.9.* If $(S; +, \times)$ has an identity, 0, for $+$ and disjoint subsets $G, L$ for which (1) $L \cup G = S - \{0\}$, (2) $G$ is closed under $+$ and $\times$, and (3) $L$ is closed under addition, then the elements of $G$ and $L$ are called *positive* and *negative* elements respectively.

From the basic relation between $F$ and $J$ it follows that those rational numbers which can be represented as $m/n$, where both $m$ and $n$ are positive integers, can be identified as positive rational numbers because, with

$$\frac{a}{b} \cdot \frac{c}{d} = \frac{ac}{bd} \quad \text{and} \quad \frac{a}{b} + \frac{c}{d} = \frac{ad + bc}{bd}$$

where $a, b, c, d \in N$, $ac, bd$ and $ad + bc$ are also positive integers. It seems quite appropriate that rational numbers whose standard fractions appear as

$$\frac{^-m}{n} = {}^-\left(\frac{m}{n}\right)$$

where $m, n \in N$ should qualify as negative numbers since $^-m/n$ is the negative of $m/n$ just as $^-m$ is the negative of $m$. These remarks support the following theorem.

## Theorem 8.15

Every rational number is exclusively either zero, positive or negative.

EXAMPLE: The numbers represented by $^-7/13$ and $433/^-45$ are negative, while $47/43$ and $^-35/^-90$ represent positive rational numbers. Since an identity element is unique, there is only one rational zero.

By referring to Definition 8.9, we can define the relation "greater than" in a general way.

160    *Elementary Mathematical Structure*

*Definition 8.10.* In a system $(S; +, \times)$ an element $b \in S$ is *greater than* $a \in S$ (written $b > a$) iff $b - a$ is positive in $(S; +, \times)$. A statement equivalent to $b > a$ is $a < b$ (read *a is less than b*).

EXAMPLE: $17 > 11$ because $17 - 11 = 6$ is a positive integer. $^-3 > ^-5$ because $^-3 - ^-5 = ^-3 + 5 = 2$ is positive.

$$\frac{^-7}{12} > \frac{^-11}{18}$$

because

$$\frac{^-7}{12} - \frac{^-11}{18} = \frac{^-21}{36} - \frac{^-22}{36} = \frac{^-21 - ^-22}{36} = \frac{^-21 + 22}{36} = \frac{1}{36}$$

which is positive.

*Exercise 8.5*

1. If $2\frac{3}{7}$ means $2 + \frac{3}{7}$, show the meaning of each of the following and write an equivalent common fraction:
   (a) $4\frac{13}{15}$   (b) $2\frac{1}{8} + 3\frac{3}{8}$   (c) $\frac{13}{14} + \frac{9}{14}$   (d) $4\frac{2}{3} + 3\frac{1}{4}$

2. Use the properties $0/n = 0$ and $n/n = 1$ to write the results of the following by sight:
   (a) $\frac{11}{15} + \frac{0}{15}$   (b) $\frac{13}{17} \times \frac{43}{43}$   (c) $\frac{213}{476} \times \frac{0}{49}$

*3. Write each of the following as a product of primes:
   (a) 264   (b) 585   (c) 3969   (d) 2261

In performing the operations for Problems 4 through 7, use prime factors to best advantage and represent each result by a standard fraction.

*4. (a) $\frac{8}{27} \times \frac{45}{28}$   (b) $\frac{18}{33} \cdot \frac{143}{30} \cdot \frac{133}{221}$   (c) $\frac{130}{231} \cdot \frac{99}{156}$

*5. (a) $\frac{5}{12} + \frac{14}{21}$   (b) $\frac{9}{91} + \frac{^-2}{39}$   (c) $\frac{5}{82} + \frac{4}{123}$

*6. (a) $\frac{9}{10} \div \frac{3}{5}$   (b) $\frac{^-8}{45} \div \frac{28}{27}$   (c) $\frac{^-69}{^-68} \div \frac{46}{^-51}$

7. (a) $\frac{2}{21} - \frac{5}{14}$   (b) $\frac{2}{39} - \frac{9}{91}$   (c) $\frac{10}{63} - \frac{9}{91} - \frac{11}{39}$

8. Identify the element of $F$ which is the (1) additive inverse; (2) multiplicative inverse of each of the following:

(a) $\dfrac{3}{-7}$  (b) $^-5$  (c) $\dfrac{0}{6}$  (d) $\dfrac{4}{6}$

9. Indicate which number of each pair is greater:

(a) $^-7, ^-3$    (d) $\dfrac{^-5}{6}, \dfrac{3}{^-4}$

(b) $2, ^-3$    (e) $\dfrac{25}{57}, \dfrac{17}{39}$

(c) $0, ^-5$    (f) $\dfrac{^-3}{100}, \dfrac{0}{^-7}$

*10. Prove that if $n$ is a negative number then $^-n$ is a positive number. (Hint: $^-n$ is either negative, zero or positive by Definition 8.9. Show that $^-n$ is neither zero nor negative.)

*11. Use the result of Problem 10 to prove that $p > 0$ and $0 > n$ for every positive number $p$ and every negative number $n$.

*12. Prove that $p > n$ for every positive number $p$ and every negative number $n$.

# 9

# Extending the Decimal Numeration System

We have emphasized the importance of an adequate numeration system, and have tried to make clear the way in which computations become easier by the interrelation of our decimal place value system and the basic properties of our number systems. With the introduction of rational numbers, we have resorted to the use of ordered *pairs* of decimal numerals to represent the new numbers, since the quotient was not always an integer. Having finished our discussion of the nature of the numbers, we turn our attention back to the problem of representing them. Is it possible to represent every rational number by a decimal numeral? If so, what modifications need to be made to our numeration system to accomplish this, and what kinds of algorithms will we have?

## EXTENDING THE PLACE VALUE CONCEPT

In the initial description of the decimal system of numeration, we said that the right-most digit of the numeral always has place value one. Physically speaking, however, there is just as much room to the right of the units digit as there is to the left. If we were to write a digit to the right of the units digit, what must its place value be in order to conform to the pattern which exists for the other places? More explicitly, how does the place value of one digit compare to that of the digit to its left? Let us examine a few familiar cases.

Place value

$$1 = \frac{10}{10} = 10 \times \frac{1}{10}$$

$$10 = \frac{100}{10} = 100 \times \frac{1}{10}$$

$$100 = \frac{1000}{10} = 1000 \times \frac{1}{10}$$

From these examples it should be clear that multiplication of a place value of a digit by 1/10 produces the place value of the digit to its right. The place value of the digit to the right of the units digit would be

$$1 \times \frac{1}{10} \quad \text{or} \quad \frac{1}{10}$$

Now that we know the place value of the digit to the right of the units digit, how do we write a digit there? For example, if we have the integer 234 and we simply write the digit 5 to the right of the digit 4, by our previous agreement we must interpret the results, 2345, as two thousand three hundred forty-five where 5 is now taken to be in units place. The traditional American solution to this problem is the so-called decimal *point* (the punctuation symbol used for a period) which is placed between the 4 and the 5, that is, immediately to the right of the units digit, as in 234.5.

With these observations we are able to extend place values as shown:

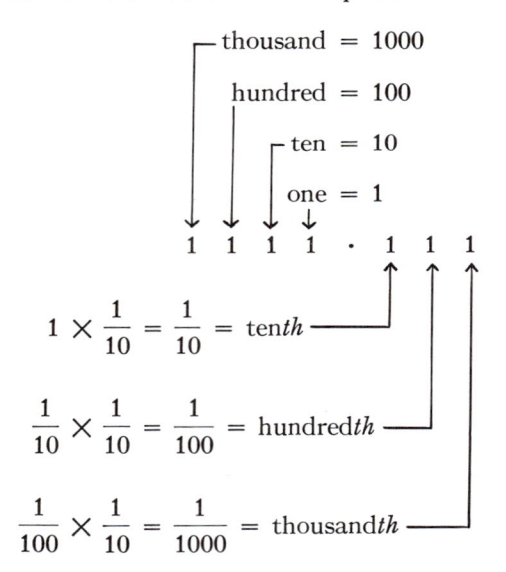

$$1 \times \frac{1}{10} = \frac{1}{10} = \text{ten}th$$

$$\frac{1}{10} \times \frac{1}{10} = \frac{1}{100} = \text{hundred}th$$

$$\frac{1}{100} \times \frac{1}{10} = \frac{1}{1000} = \text{thousand}th$$

where the italicized *th* emphasizes the similarity in the words used to describe the place value of the pair of digits which occupy corresponding positions on opposite sides of the units digit. For example, the 2 and 3 in the decimal numeral

$$121110.555314$$

are each separated from 0, the units digit, by three other digits and 2 has place value *ten thousand* while 3 has place value *one ten thousandth*, or 10,000 in one case and 1/10,000 in the other. You can see that great care must be taken not to confuse such place values either in spelling or in pronouncing them. As

*Extending the Decimal Numeration System* **165**

with whole numbers, place values are represented by the digit 1 in the place represented, with zeros in the other places:

$$.1 = \frac{1}{10}$$

$$.01 = \frac{1}{100}$$

$$.001 = \frac{1}{1000}$$

$$.0001 = \frac{1}{10,000}$$

EXAMPLE: We will use the numeral 325.746 as an example of decimal values. In our algorithms we might expect the use of expanded form, since it was used before. In this form, 325.746 =

$$3 \times 100 + 2 \times 10 + 5 \times 1 + 7 \times \frac{1}{10} + 4 \times \frac{1}{100} + 6 \times \frac{1}{1000}$$

or

$$300 + 20 + 5 + \frac{7}{10} + \frac{4}{100} + \frac{6}{1000}$$

which is a rational number since $F$ is closed under addition. Expressed in words, the numeral is separated into $325 + .746$ and the fractional part is read according to the following result:

$$.746 = \frac{7}{10} + \frac{4}{100} + \frac{6}{1000}$$

$$= \frac{7 \times 100}{10 \times 100} + \frac{4 \times 10}{100 \times 10} + \frac{6}{1000}$$

$$= \frac{7 \times 100 + 4 \times 10 + 6}{1000}$$

$$= \frac{746}{1000}$$

Thus 325.746 is read *Three hundred twenty-five and seven hundred forty-six thousandths.*

## ADDITION AND SUBTRACTION

Since we have used the same relative pattern of place value in our extended decimal numeration system as in the earlier system, we can expect the same general patterns of digit-by-digit operations in the algorithms. In addition and subtraction, digits of the same place value will be combined. This is shown in the following example by the use of the expanded form and the basic properties of $(F; +, \times)$.

EXAMPLE:

$$21.3 + 0.485 = \left(2 \times 10 + 1 + 3 \times \frac{1}{10}\right)$$

$$+ \left(4 \times \frac{1}{10} + 8 \times \frac{1}{100} + 5 \times \frac{1}{1000}\right)$$

$$= 2 \times 10 + 1 + (3 + 4) \times \frac{1}{10} + 8 \times \frac{1}{100} + 5 \times \frac{1}{1000}$$

$$= 2 \times 10 + 1 + 7 \times \frac{1}{10} + 8 \times \frac{1}{100} + 5 \times \frac{1}{1000}$$

$$= 21.785$$

It is customary, but not mandatory, to place a zero in units place when the number is between $-1$ and $1$, as in 0.485. If several numbers are to be added, it is helpful to write the numerals in a column with the digits of the same place value also in a column. Thus $21.3 + 0.485$ could have been written as

$$\begin{array}{r} 21.3 \\ 0.485 \\ \hline 21.785 \end{array}$$

where there is less danger of combining digits of different place values.

## MULTIPLICATION

The multiplication algorithm can be examined from the expanded form of numeration also. There is another viewpoint, however, which takes advantage of previous experience with the multiplication algorithm of the

## Extending the Decimal Numeration System

system $(W; +, \times)$, so we will develop and illustrate the procedure with an example.

EXAMPLE: Since $2.35 = \dfrac{235}{100}$ and $4.6 = \dfrac{46}{10}$

$$2.35 \times 4.6 = \dfrac{235}{100} \times \dfrac{46}{10}$$

$$= \dfrac{235 \times 46}{1000}$$

This procedure indicates the possibility of performing the digit-by-digit multiplication as if the final digit of each factor had place value one; that is, we ignore the decimal point for this part of the procedure and obtain

$$\dfrac{10,810}{1,000} = 10.810$$

The location of the decimal point can be ascertained by observing that

$$2 < 2.35 < 3$$

and

$$4 < 4.6 < 5,$$

so that

$$2 \times 4 < 2.35 \times 4.6 < 3 \times 5$$

or the *whole number part* of the number must be 10 and not 1 or 108. It is more economical in time and effort to note that the place value of the last digit, 0, of the product is the product of the place values of the last digits, 5 and 6, of the factors, namely

$$\dfrac{1}{100} \times \dfrac{1}{10} = \dfrac{1}{1000}$$

This observation is directly related to the number of digits to the right of the decimal point in each case. Thus the customary algorithm is written as

$$\begin{array}{r} 2.35 \\ 4.6 \\ \hline 1410 \\ 940 \\ \hline 10.810 \end{array}$$

where the decimal point in the product was placed so that there were three digits to its right since 2.35 has two and 4.6 has one.

168                Elementary Mathematical Structure

## DIVISION

An algorithm can be devised for division that is simply an extension of that used in $(W; +, \times)$. The use of the principle expressed in Problem 10(a), Exercise 7.2 and Theorem 8.9 provides a way to simplify the algorithm.

EXAMPLE:

$$10.81 \div 4.6 = \frac{10.81}{4.6}$$

$$= \frac{10.81 \times 10}{4.6 \times 10}$$

$$= \frac{108.1}{46}$$

This procedure makes it possible to perform division in many cases by using a natural number for a divisor:

```
         2.35
    46|108.10
       92 00
       -----
       16 10
       13 80
       -----
        2 30
        2 30
       -----
```

Two things should be observed in concluding this example: first, $108.1 = 108.10$, and second, when digits are carefully placed and the divisor is a natural number, each digit of the quotient has the same place value as the digit in the dividend directly below it, with the decimal point of the quotient located immediately above that of the dividend.

*Exercise 9.1*

  1. Write in expanded form
     (a) 234.56    (b) 50.004    (c) 0.00307
  *2. Write word symbols for each numeral in Problem 1.
  3. Use the extended place value concept to write expanded form numerals for the following:
     (a) $32.421_{six}$    (b) $1101.0101_{two}$    (c) $0.0123_{four}$
  *4. Give the place value in both fraction and word form for the digit 3 in each of the following:
     (a) $10.31_{seven}$    (b) $0.103_{six}$    (c) $0.0010230_{four}$

*Extending the Decimal Numeration System*             **169**

**5.** Use expanded form to add the following:
(a) $20.13 + 7.254$     (b) $1.007 + 0.8391$

**6.** Perform the following according to the usual algorithm:
(a) $20.13 + 7.254$     (b) $20.134 - 7.25$     (c) $1.007 - 0.8391$

***7.** Write each of the following as a common fraction:
(a) $2.34$     (b) $0.73$     (c) $1.007$     (d) $0.0305$

***8.** Write a decimal numeral for each of the following:

(a) $\dfrac{403}{10{,}000}$     (b) $\dfrac{30756}{100}$     (c) $\dfrac{47}{20}$     (d) $\dfrac{33}{2 \cdot 2 \cdot 2 \cdot 5}$

**9.** Perform the indicated multiplications, after expressing the numbers in common fraction form. After multiplying, convert the product numeral to decimal form.
(a) $20.3 \times 2.7$     (b) $0.023 \times 4.07$     (c) $0.105 \times 0.006$

**10.** Perform the multiplications of Problem 9 by the usual method (algorithm).

***11.** Express the following as quotients of whole numbers:

(a) $107.041 \div 4.07$     (b) $43 : 0.017$     (c) $\dfrac{0.006}{0.0123}$

***12.** Perform the indicated divisions, after converting to an equivalent form where the divisor is a whole number. (See the example preceding Exercise 9.1.)
(a) $10.7041 \div 4.07$     (b) $1070.41 \div 0.407$

**13.** Perform the indicated divisions with the numerals unchanged:
(a) $107.041 \div 4.07$     (b) $0.07982 \div 0.026$

**14.** Use the inverse relation between $\times$ and $\div$ to devise a rule for placing the decimal point in the quotient.

## REPEATING DECIMALS

You may have noticed in the preceding exercise that the only type of common fraction to be converted to a decimal was one which had only 2 and 5 as prime factors of its denominator. In such a case, enough factors of 2 or 5 can be introduced to produce an equivalent fraction whose denominator is a product of tens:

$$\frac{33}{40} = \frac{33}{2 \cdot 2 \cdot 2 \cdot 5} = \frac{33 \cdot (5 \cdot 5)}{2 \cdot 2 \cdot 2 \cdot 5(5 \cdot 5)}$$

$$= \frac{33 \cdot 25}{(2 \cdot 5)(2 \cdot 5)(2 \cdot 5)} = \frac{825}{1000}$$

$$= 0.825$$

Such a number has a *terminating* decimal numeral. The significance of this expression will become clearer after we examine a different type fraction. First, however, you should recognize that the result of the problem could have been obtained by using the division algorithm.

$$\begin{array}{r} .825\phantom{0} \\ 4\overline{\smash{)}3.300} \\ 3\phantom{.}200\phantom{0} \\ \hline 100\phantom{0} \\ 80\phantom{0} \\ \hline 20\phantom{0} \\ 20\phantom{0} \\ \hline \end{array}$$

Notice that any number of zeros placed before or after the digits of a numeral *with decimal point in place* produces a numeral for exactly the same number. Thus 00003.3 = 3.3000 = 003.30000 = 3.3. Of course, the zeros on the left are useless, but those to the right are helpful in performing the division algorithm.

In order to learn the nature of the decimal numeral for a fraction having prime factors other than 2 and 5 in its denominator, we will examine two such fractions having the least possible prime denominators. In the first case,

$$\begin{array}{r} 0.333 \\ 3\overline{\smash{)}1.000} \\ 900 \\ \hline 100 \\ 90 \\ \hline 10 \\ 9 \\ \hline 1 \end{array} \qquad \begin{array}{r} 0.142857 \\ 7\overline{\smash{)}1.000000} \\ 7000 \\ \hline 3000 \\ 2800 \\ \hline 200 \\ 140 \\ \hline 60 \\ 56 \\ \hline 40 \\ 35 \\ \hline 50 \\ 49 \\ \hline 1 \end{array}$$

1/3, we see that after each division we are faced again with the same basic quotient, 10 ÷ 3, so although we stopped the process, the division is not completed. If we look at the partial quotient at the end of each step we see the following sequence:

$$0.3, \ 0.33, \ 0.333, \ 0.3333, \ 0.33333, \ \cdots$$

where the three dots indicate that there is no end to the list. Thus we find another situation in which suspension points are useful. Mathematicians write

$$\frac{1}{3} = 0.333\cdots$$

to mean that there is no end to the sequence of 3's.

The case of 1/7 is somewhat different, in that a new remainder appeared with each division, so in sequence, 7 had to divide 10, 30, 20, 60, 40, and 50 before 10 occurred again. Notice that a remainder of zero would have occurred only if one of these dividends had been a multiple of 7. But these numbers are all of the form $n \times 10$ where $0 < n < 7$, and 7 is neither a factor of $n$ nor of 10, so it is impossible for 7 to divide them. Observed also that in the case of division by 7, the only possible remainders are the six numbers 1, 2, 3, 4, 5, and 6, all of which occurred. Thus, after six divisions at the most, there had to be a repetition of one of these remainders. Then the same sequence of remainders (hence the same sequence of digits in the quotient) starts anew, so that

$$\frac{1}{7} = 0.142857142857\cdots = 0.\overline{142857}$$

where the bar above 142857 indicates endless repetition of this sequence of digits just as the three dots do. Notice that the use of suspension points requires that the sequence be written *twice* to identify the digits. For example,

$$\frac{100}{7} = 14.2857142857\cdots$$

while

$$\frac{1}{70} = 0.0142857142857\cdots$$

From a broader point of view, the cases of 1/3 and 1/7 are not different. In both cases the decimal numeral has a sequence of digits which is repeated endlessly. In the case of 1/3 it consists of a single digit, while in the case of 1/7 it is composed of six digits. From this viewpoint we may also observe that a number like 33/40 has the same characteristic, in that

$$\frac{33}{40} = 0.825 = 0.825000\cdots$$

where the digit zero repeats endlessly. From these observations it is a reasonable conclusion that *every rational number has a repeating decimal representation*. It would be convenient to have a unique decimal representation, but this is not always the case. We will demonstrate this fact in the following example and deal with it further in the succeeding development.

EXAMPLE: We have demonstrated that

$$\frac{1}{3} = 0.333\cdots$$

in the sense that the numbers in the sequence

$$0.3, \quad 0.33, \quad 0.333, \quad 0.3333, \quad 0.33333, \quad \cdots$$

become increasingly and arbitrarily close to the number 1/3. In other words, these numbers differ from 1/3 by less than

$$0.1, \quad 0.01, \quad 0.001, \quad 0.0001, \quad 0.00001, \quad \cdots$$

respectively. By the same token, the numbers in the sequence

$$0.9, \quad 0.99, \quad 0.999, \quad 0.9999, \quad 0.99999, \quad \cdots$$

become increasingly and arbitrarily close to the number $3 \times 1/3 = 1$. That is,

$$1 = 3 \times 0.333\cdots = 0.999\cdots$$

whereas we are more accustomed to the idea that

$$1 = 1.000\cdots$$

Actually, it turns out that every terminating decimal has *two* endlessly repeating representations, one which employs zeros and another which employs nines. As another example,

$$\frac{33}{40} = 0.825 = 0.8249999\cdots.$$

If every rational number has a repeating decimal representation, the possibility of the converse situation becomes interesting. That is, does every repeating decimal represent a rational number? For example, does

$$0.123123123\cdots$$

represent a rational number? For the present, our only way to identify a rational number is to determine whether it can be represented in common fraction form, which involves two whole numbers. Multiplication of $0.123123\cdots$ by powers of ten will produce numerals with digits to the left of the decimal point such, as

$$1.231231\cdots$$

$$12.312312\cdots$$

$$123.123123\cdots$$

The last of these numerals seems best related to the numeral

$$0.123123\cdots$$

since both have exactly the same repeating sequence after the decimal point. With this observation, the following calculations and relations should seem both natural and meaningful,

$$123.123123\cdots - 0.123123\cdots = 123$$

or

$$1000 \times 0.123123\cdots - 1 \times 0.123123\cdots = 123$$

or

$$(1000 - 1) \times 0.123123\cdots = 123$$

that is

$$999 \times 0.123123\cdots = 123$$

so that

$$0.123123\cdots = \frac{123}{999} = \frac{41}{333}$$

which certainly represents a rational number.

From this example you should understand that multiplying a repeating decimal having $n$ digits in its repeating sequence by $10^n$ or $n$ tens produces a numeral which has the same repeating sequence. The difference of the numbers has repeating zeros, and so is a terminating decimal. It is also the product of

$$10^n - 1 \quad \text{or} \quad 99\cdots9$$

and the original number. The examples

$$0.333\cdots$$

and

$$0.999\cdots$$

furnish interesting cases.

$$3.33\cdots - 0.33\cdots = 3$$
$$10 \times 0.33\cdots - 1 \times 0.33\cdots = 3$$
$$9 \times 0.33\cdots = 3$$
$$0.33\cdots = 3 \div 9 = \frac{1}{3}$$

$$9.99\cdots - 0.99\cdots = 9$$
$$10 \times 0.99\cdots - 1 \times 0.99\cdots = 9$$
$$9 \times 0.99\cdots = 9$$
$$0.99\cdots = 9 \div 9 = 1$$

Although we have ignored the negative rationals and have merely illustrated the principles involved, you should find the following statement plausible. It is listed as a theorem, although no formal proof is proposed.

## Theorem 9.1

Every rational number has a repeating decimal representation, and every repeating decimal numeral represents a rational number.

# DECIMAL NUMERALS AND THE NUMBER LINE

According to the location of points represented by common fractions, the equally spaced points pictured below between $a$ and 1 represent the numbers

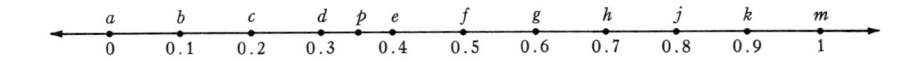

$$0.1, 0.2, 0.3, 0.4, 0.5, 0.6, 0.7, 0.8, 0.9$$

```
    a     b     c     d  p   e     f     g     h     j     k     m
    •─────•─────•─────•──•───•─────•─────•─────•─────•─────•─────•
    0    0.1   0.2   0.3    0.4   0.5   0.6   0.7   0.8   0.9    1
```

The left end point of each of the ten sub-segments of $\overline{am}$ is designated by

$$0.n$$

where $n$ is a digit. Now, if $p$ is a point on $\overrightarrow{ab}$ for which

$$|am| = 3|ap|$$

then $p$ represents the number $1/3$. We will show that

$$d(p)e \quad \text{and} \quad |de| = 3|dp|$$

Since the coordinates of $d$, $p$ and $e$ are $0.3$, $1/3$, and $0.4$ respectively and since

$$0.3 < \frac{1}{3} < 0.4$$

it follows by the order of points on the number line that

$$|ae| > |ap| > |ad|$$

so that

$$d(p)e$$

We also have, by the construction of the number line,

$$|ad| = 3|ab| = |eh| = |hm|$$

so that

$$3|ad| = |ad| + |eh| + |hm|$$

whence

$$3|ap| = 3(|ad| + |dp|) = 3|ad| + 3|dp|$$
$$= |ad| + |eh| + |hm| + 3|dp|$$

But,

$$3|ap| = |am| = |ad| + |eh| + |hm| + |de|$$

so that
$$|de| = 3|dp|$$
Thus $p$ is related to $d$ and $e$ by position, just as it was with respect to $a$ and $m$. Subdividing $\overline{de}$ into ten congruent sub-intervals by points with coordinates

$$0.31, 0.32, 0.33, 0.34, 0.35, 0.36, 0.37, 0.38, 0.39$$

locates $p$ between the points with coordinates

$$0.33 \quad \text{and} \quad 0.34$$

in the same relative position as before. A continuation of this procedure gives a sequence of inequalities

$$0 < \frac{1}{3} < 1$$

$$0.3 < \frac{1}{3} < 0.4$$

$$0.33 < \frac{1}{3} < 0.34$$

$$0.333 < \frac{1}{3} < 0.334$$

$$\vdots$$

with the sequence of coordinates of the points between $a$ and $p$, namely

$$0, \ 0.3, \ 0.33, \ 0.333, \ \cdots$$

furnishing the decimal representation for the point $p$, so that

$$0.333\cdots = \frac{1}{3}$$

as before.

## Exercise 9.2

*1.  Write a decimal numeral for each of the following:

(a) $\dfrac{8}{11}$  (b) $\dfrac{8}{3}$  (c) $\dfrac{8}{33}$  (d) $\dfrac{8}{7}$  (e) $\dfrac{13}{21}$  (f) $\dfrac{3}{17}$

*2. Write each of the following as a non-terminating decimal numeral in *two* ways:

(a) 41.7  (b) $\dfrac{17}{20}$  (c) 20  (d) 0.0015

*3. Write a common fraction to represent each of the following:
(a) $0.666\cdots$  (b) $0.2727\cdots$  (c) $0.\overline{621}$  (d) $-3.693693\cdots$

4. Write a decimal numeral for each of the following:
(a) $0.101_{two}$  (b) $2.03_{four}$  (c) $0.23_{six}$  (d) $0.1313_{five}$

5. Since

$$\frac{1}{3} = \frac{8 \cdot \frac{1}{3}}{8} = \frac{2\frac{2}{3}}{8} = \frac{2}{8} + \frac{\frac{2}{3}}{8} = \frac{2}{8} + \frac{8 \cdot \frac{2}{3}}{8^2}$$

$$= \frac{2}{8} + \frac{5\frac{1}{3}}{8^2} = \frac{2}{8} + \frac{5}{8^2} + \frac{\frac{1}{3}}{8^2} = \frac{2}{8} + \frac{5}{8^2} + \frac{1}{8^2}\left(\frac{1}{3}\right)$$

$$= \frac{2}{8} + \frac{5}{8^2} + \frac{2}{8^3} + \frac{5}{8^4} + \cdots$$

we see that $\frac{1}{3}$ can be expressed in base eight as $0.2525\cdots_{eight}$. Find similar numerals for $\frac{1}{6}$ in base (a) eight (b) nine (c) three (d) two.

6. Use long division in the base indicated to find the numerals of Problem 5. The work for the example is shown below.

$$\begin{array}{r} .2525\cdots \\ 3_{eight}\overline{\smash{\big)}1.00_{eight}} \\ \underline{60} \\ 20 \\ \underline{17} \\ 1 \end{array}$$

7. Use the principles of this section to find a common fraction in the indicated base for the following: (a) $0.\overline{1313}\cdots_{four}$  (b) $0.333\cdots_{six}$  (c) $0.77\cdots_{eight}$  (d) $0.\overline{101}_{two}$.

8. Write a decimal numeral for each of the numbers of Problem 7.

*9. On your own paper, duplicate the picture below. Subdivide the segments containing point $p$ into halves, fourths, eighths, etc., to find the first four digits to the right of the decimal point of the base two numeral for 1/5 (the coordinate of $p$). Use methods of Problems 5 or 6 to check your result.

10. Find a way to perform the following operations without using the common fraction forms.
 (a) $0.123123\cdots + 1.231231\cdots$   (b) $1.231231\cdots - 0.123123\cdots$
 (c) $9 \times 0.123123\cdots$   (d) $0.7272\cdots \div 0.66\cdots$

*11. Check your results in Problem 10 using the common fraction forms.

## IRRATIONAL NUMBERS

Although many interesting and practical things can be accomplished within the system of whole numbers, we have found it inadequate in many situations. A typical situation which we have used to illustrate this inadequacy is the search for solution sets of equations. For example, the equation

$$x + 13 = 6$$

points up the need for negative numbers, while

$$6x = 13$$

shows the need for fractions. It may seem surprising, but the simple equation

$$n \cdot n = 2$$

or

$$n^2 = 2$$

as we commonly write it, does not have a solution in the set $F$ of rational numbers. As an implication, this idea can be stated as follows: If $n \in F$, then $n^2 \neq 2$. We shall make an indirect proof by using the observation that an implication $p \rightarrow q$ is false only when $p$ is true and $q$ is false. In other words, we will see that it is impossible for $n \in F$ to be true when $n^2 \neq 2$ is false; that is, $n \in F$ and $n^2 = 2$ cannot both be true.

Suppose $n \in F$ and $n^2 = 2$ *are* both true. (Incidentally, if there *is* such a rational number, $r$, and it is negative then $^-r$ is a positive rational and also a solution, so we may just as well confine our discussion to positive numbers.) $n \in F$ and $n > 0$ imply that there are natural numbers $p$ and $q$ for which

$$n = \frac{p}{q}$$

and

$$n^2 = \frac{p}{q} \cdot \frac{p}{q} = p^2 \div q^2$$

Then, since we are also assuming that $n^2 = 2$,

$$p^2 \div q^2 = 2$$

or

$$p^2 = 2 \cdot q^2$$

Since $p$ and $q$ are natural numbers, each is the product of a unique set of prime

numbers; that is
$$p = p_1 p_2 \cdots p_r$$
$$q = q_1 q_2 \cdots q_s$$
where $p_1, p_2, \cdots, p_r, q_1, q_2, \cdots, q_s$ are prime numbers. Then
$$p^2 = p_1 p_1 p_2 p_2 \cdots p_r p_r = 2 q_1 q_1 q_2 q_2 \cdots q_s q_s = 2q^2$$
so that
$$p_1 p_1 p_2 p_2 \cdots p_r p_r$$
and
$$2 q_1 q_1 q_2 q_2 \cdots q_s q_s$$
both represent the same number. But whereas
$$p_1 p_1 p_2 p_2 \cdots p_r p_r$$
is the product of an *even* number $(2 \cdot r)$ of prime factors, the number
$$2 q_1 q_1 q_2 q_2 \cdots q_s q_s$$
is the product of an *odd* number $(2 \cdot s + 1$, since the factor 2 is also a prime) of prime factors, so the two sets of primes are not the same and therefore do *not* represent the same number. Thus the hypothesis that $n \in F$ and $n^2 = 2$ are both true implies the false conclusion that two expressions are both equal and unequal. Since only a false hypothesis can imply a false conclusion, the hypothesis that both $n^2 = 2$ and $n \in F$ must be false, as we predicted. Therefore there is no rational number which multiplies by itself to produce 2.

Since there is no element $n \in F$ for which $n^2 = 2$, we may well ask, "Is there any *other* kind of number whose square is 2?." Perhaps the best way to answer this question is to present a practical situation where the number makes a rather natural appearance. In the illustration, if the measure of the lengths of the sides of square $\overline{abcd}$ is 1, then the measure, $x$, of the length of the

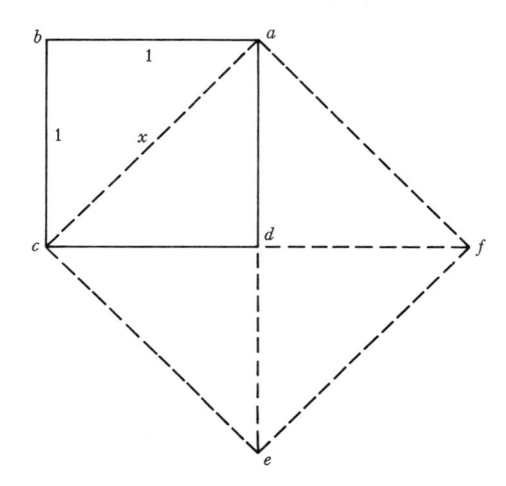

### Extending the Decimal Numeration System

sides of square $\overline{acef}$ is such that
$$x^2 = 2$$
since the interior of $\overline{acef}$ is composed of 4 triangular regions congruent to the triangular region $\overline{abc}$ and the measure of the area of a square is found by squaring the measure of the length of its side.

Because of the need for solutions to certain algebraic equations and the need for certain measures of length in geometry, there is ample reason to define other numbers which can be represented by points on the number line. Those which are not rational are called *irrational* and the set composed of the union of the set of rationals and the set of irrationals is called the set of *real* numbers. Numbers like $\sqrt{2}$ (*square root of 2*, the positive solution of $x^2 = 2$) which are solutions of equations of the form

$$a_n x^n + a_{n-1} x^{n-1} + \cdots + a_1 x + a_0 = 0$$

where $a_0, a_1, \cdots, a_n$ are integers, are called *algebraic* numbers. Thus, the set of rational numbers is a subset of the set of algebraic numbers. Real numbers which are not algebraic are called *transcendental* numbers. The familiar symbol $\pi$ which represents the measure of the circumference of a circle whose diameter is of measure 1, represents a transcendental number. The inherent difficulty of establishing these distinctions concerning real numbers is demonstrated by the fact that although the number $\pi$ was recognized as early as 500 B.C., the first proof that it is transcendental was not published until 1882.

The following diagram will illustrate the imbedding or extension relationships between the various classifications of numbers we have mentioned.

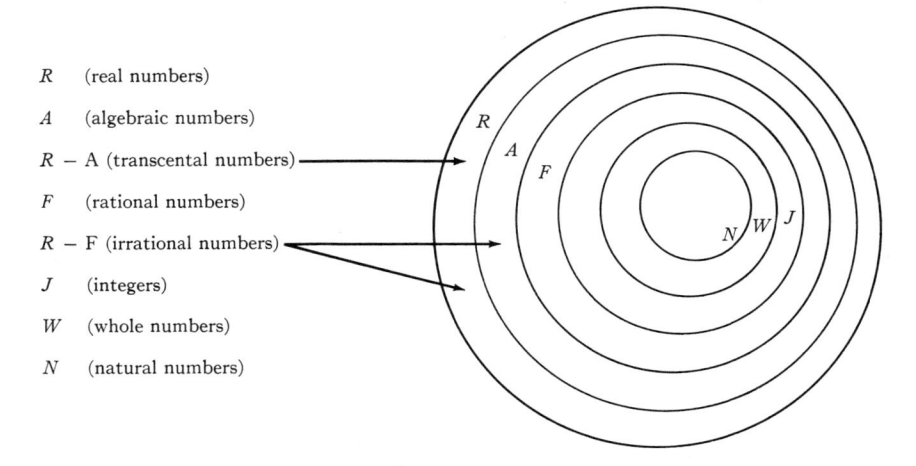

R    (real numbers)
A    (algebraic numbers)
R − A (transcental numbers)
F    (rational numbers)
R − F (irrational numbers)
J    (integers)
W    (whole numbers)
N    (natural numbers)

One of the logically sound presentations of the system of real numbers is its development as an extension of the system $(F; +, \times)$ as a result of the work of the German mathematician Georg Cantor (1845–1918). In an oversimplified version, his approach runs somewhat as follows. We know that some numbers can be defined by sequences of rational numbers in the manner that the sequence

$$.3, .33, .333, .3333, \cdots$$

defines the number one-third. Such a non-ending, or infinite, sequence is said to *converge* to the number. Cantor called these *fundamental* sequences. Would it not be possible for such an infinite sequence of rational numbers to converge to something which is not a rational number? The difficulty of establishing an affirmative answer to this question lies in the method of deciding whether or not a given sequence converges. One method is to establish the fact that no matter how small a positive rational number, $p$, that is selected, all of the numbers of the sequence which come after a certain one (depending on $p$) will differ from the number in question by a positive amount which is less than $p$. For example, if we choose

$$p = 0.00001$$

then all the numbers

$$.33333, .333333, .3333333, \cdots$$

differ from 1/3 by less than 0.00001, as indicated by the difference of 1/3 and the first number,

$$\frac{1}{3} - .33333 = \frac{1}{3} - \frac{33{,}333}{100{,}000}$$

$$= \frac{100{,}000}{300{,}000} - \frac{99{,}999}{300{,}000} = \frac{1}{300{,}000} = \frac{1}{3} \times 0.00001$$

which is certainly less than 0.00001.

The disadvantage of this method as a *definition* of a new kind of number is that the number itself must be identified to determine whether the sequence in question actually converges so as to define any number. Fortunately, there is another method, initiated in Euclidean geometry, and placed on a rigorous footing by the prolific French mathematician Augustin-Louis Cauchy (1789–1857) by which the convergence of an infinite sequence can be established without any knowledge of or reference to the number to which it converges. According to the Cauchy criterion, for an infinite sequence to converge, a necessary and sufficient condition is that for every rational $p > 0$ all numbers in the sequence which occur after a certain one (depending on $p$) shall differ *from each other by a positive amount which is less than $p$*. This criterion has had such an impact on mathematics, that many people call a convergent sequence a *Cauchy sequence.*

By means of the Cauchy criterion for sequence convergence, Cantor was able to show the existence of sequences of rational numbers which converge but do not converge to any *rational* number. It was also possible to determine whether two different sequences "converge to the same number," so to speak. Such a *class* of sequences was taken as the definition of the number, and such numbers were called *irrational*, while the set of *all* numbers defined by classes of fundamental sequences of rational numbers was called the set of *real* numbers, $R$. Cantor was also able to prove that the set of real numbers is *complete*, whereas the set $F$ is not, in that equivalence classes of fundamental sequences of real numbers define only real numbers, so that $R$ cannot be extended by this technique. From the geometric point of view this means that *all* points on a line with a coordinate system have *real* coordinates; that is, the points of a number line which have rational coordinates together with those whose coordinates are irrational constitute the entire line.

## DECIMAL REPRESENTATION OF REAL NUMBER

The development described in the preceding section is the work of a mathematical genius and, as such, is certainly not contemplated for use in this presentation. We will, however, use the basic ideas as a guide for the development which we propose. Instead of considering all equivalent sequences of rational numbers, we will consider only a single sequence of a particular type. In the light of Cantor's argument, it should seem plausible that

$$n.d_1 \leq n.d_1d_2 \leq n.d_1d_2d_3 \leq \cdots$$

(where $d_1, d_2, d_3, \cdots$ are digits and $n \in W$) is a sequence which converges to a number $m$ for which

$$n.d_1 \leq m < n.d_1 + 0.1$$
$$n.d_1d_2 \leq m < n.d_1d_2 + 0.01$$
$$n.d_1d_2d_3 \leq m < n.d_1d_2d_3 + 0.001$$

so that the unending decimal

$$n.d_1d_2d_3\cdots$$

defines the number $m$.

### Axiom 9.1

There is a set $R$ of *real* numbers each of which is represented by a non-ending decimal $n.d_1d_2\cdots$ (or $^-n.d_1d_2\cdots$) where $n$ is a whole number decimal and $d_1, d_2, \cdots$ are digits.

The following theorem is proved in more advanced courses by methods related to Cantor's work, but we will simply state it here to indicate the direction of development.

## Theorem 9.2

If $m.b_1b_2\cdots$ and $n.d_1d_2\cdots$ represent real numbers then the sequences

$$m + n,\ m.b_1 + n.d_1,\ m.b_1b_2 + n.d_1d_2,\ \cdots$$

and

$$m \times n,\ m.b_1 \times n.d_1,\ m \cdot b_1b_2 \times n.d_1d_2,\ \cdots$$

both converge.

By Axiom 9.1, the sequences of sums and products of Theorem 9.2 converge to real numbers, permitting the following definition:

*Definition 9.1.* If $x = m.b_1b_2\cdots$ and $y = n.d_1d_2\cdots$ then $x \oplus y$ is the real number to which the sequence $m + n,\ m.b_1 + n.d_1,\ m.b_1b_2 + n.d_1d_2,\ \cdots$ converges, while $x \otimes y$ represents the number to which the sequence $m \times n$, $m.b_1 \times n.d_1,\ \cdots$ converges.

Subtraction and division can be similarly defined, as can inverse elements for $\oplus$ and $\otimes$. However, the plain fact about the irrational numbers is that they are used primarily in theoretical investigations of real numbers, except in certain special cases. Therefore, we will conclude our development with a list of the properties of $(R; +, \times)$ and some exercises to help you see that the basic operations of arithmetic can be performed with real numbers in the sense of Definition 9.1.

EXAMPLE: It is known that $\pi = 3.14159\cdots$ and $\sqrt{2} = 1.41421\cdots$ From the sequences

$$3,\ 3.1,\ 3.14,\ 3.141,\ 3.1415,\ 3.14159,\ \cdots$$

and

$$1,\ 1.4,\ 1.41,\ 1.414,\ 1.4142,\ 1.41421,\ \cdots$$

the sum $\pi + \sqrt{2}$ is defined by

$$4,\ 4.5,\ 4.55,\ 4.555,\ 4.5557,\ 4.55580,\ \cdots$$

and can be expressed by $\pi + \sqrt{2} = 4.5558\cdots$ where the desired number of digits can be ascertained from the decimal numerals for $\pi$ and $\sqrt{2}$.

As a second example, we will verify that

$$\sqrt{2} = 1.41421\cdots$$

*Extending the Decimal Numeration System* **183**

by using Definition 9.1 to show that

$$1.4 < \sqrt{2} < 1.5$$
$$1.41 < \sqrt{2} < 1.42$$
$$1.414 < \sqrt{2} < 1.415$$
$$1.4142 < \sqrt{2} < 1.4143$$
$$1.41421 < \sqrt{2} < 1.41422$$
$$\cdot$$
$$\cdot$$
$$\cdot$$

from the principle that if $b$ and $c$ are positive, then

$$0 < b < c \quad \text{if} \quad b^2 < c^2$$

Since $(\sqrt{2})^2 = 2$ by definition, we only need to show that the square of each left rational number is less than 2 while the square of each right rational number is greater than 2. The truth of the last inequality verifies the earlier ones.

```
        1.41421                    1.41422
     ×  1.41421                 ×  1.41422
     ──────────                 ──────────
        141421                     282844
       282842                     282844
      565684                     565688
     141421                     141422
    565684                     565688
   141421                     141422
   ──────────                 ──────────
   1.9999899241               2.0000182084
```

## PROPERTIES OF $(R; +, \times)$

From the properties of $(F; +, \times)$ and Definition 9.1 it should not be surprising that the system of real numbers has the following properties.

1. $R$ is closed under $+$ and $\times$.
2. $+$ and $\times$ are commutative operations.
3. $+$ and $\times$ are associative operations.
4. $\times$ is distributive over $+$.
5. 0 and 1 are identity elements for $+$ and $\times$ respectively.
6. For every $b \in R$ there is a unique $b \in R$ for which $b + {}^-b = 0$; if $b \neq 0$ there is also a unique $\bar{b} \in R$ such that $b \times \bar{b} = 1$.
7. If $b, c \in R$ either $b < c$, $b = c$ or $c < b$ exclusively.
8. If $b, c \in R$ with $0 < b$ and $0 < c$ then $0 < b + c$ and $0 < bc$.

184        *Elementary Mathematical Structure*

Properties 1–6 identify $(R; +, \times)$ as an algebraic system called a *field*. The additional properties 7 and 8 qualify it as an *ordered* field. You should observe that the system $(F; +, \times)$ is also an ordered field.

## Exercise 9.3

1. By an argument similar to that in the text concerning $\sqrt{2}$, show that if $n^2 = 3$ then $n$ is not a rational number. For what kinds of natural numbers, $b$, can this argument be applied concerning $n^2 = b$?

*2. Which of the following represent rational and which irrational numbers?
(a) $\sqrt{9}$  (b) $3.10202\cdots$  (c) $3.1010010001\cdots$, where there is one more zero in each sequence than in the last  (d) $\sqrt{13}$  (e) $\sqrt{0.111\cdots}$

3. Use the patterns of digits to add:
(a) $.1212\cdots + .777\cdots$  (b) $.1212\cdots + .345345\cdots$
(c) $.\overline{3021} + .\overline{285401}$  (d) $^-1.1010010001\cdots + 2.000\cdots$
(e) $.1010010001\cdots + .0101101110\cdots$

*4. Is the set of irrational numbers closed under addition? Multiplication? Give arguments to substantiate your answers.

*5. If $\sqrt{2} = 1.4142\cdots$ and $\sqrt{3} = 1.7321\cdots$ give the sequence of rational numbers which defines
(a) $\sqrt{2} + \sqrt{3}$   (b) $\sqrt{2} \times \sqrt{3}$

6. (a) Is the negative of a rational number rational?
(b) Is the negative of an irrational number rational? Substantiate your answers.

7. Answer the questions of Problem 6 with the word *reciprocal* replacing the word *negative*.

8. Is the sum of a rational number and an irrational number rational? Give argument.

9. Is the product of a rational number and an irrational number ever rational? Give argument.

10. (a) Show that the reciprocal of $0.333\cdots$ is 3 by calculating mixed numbers for the sequence

$$\frac{1}{.3}, \frac{1}{.33}, \frac{1}{.333} \cdots .$$

(b) Find a few terms of the sequence which defines

$$\frac{1}{\sqrt{2}} = \frac{1}{1.4142\cdots}$$

and verify that it agrees with

$$\frac{\sqrt{2}}{2} = \frac{1.4142\cdots}{2} = 0.7071\cdots$$

# 10

# Special Techniques and Applications

Perhaps every field of learning has both its *pure* and *applied* aspects. *Pure mathematics* is the subject matter of mathematics alone, and is concerned with the nature of mathematics without regard to its uses or applications. This has been the primary viewpoint in our presentation, and was the viewpoint in the classical development by the ancient Greek culture. At the same time, however, there have always been people who could see mathematics as a tool, as a means for solving problems of a less academic nature. The ancient Egyptian engineers demonstrated knowledge of many of the properties of geometry centuries before geometry became a logically organized science. Even the great Archimedes (287–212 B.C.), who contributed so much to pure mathematics, used it to develop physical theories, to design mechanisms and engines of war at the command of his king. The applied mathematician is very much interested in the ways in which mathematics can be used to solve practical problems. These two viewpoints, though opposed in purpose and philosophy, have complemented each other in the development of mathematical ideas.

In most elementary studies of mathematics the student is shown both sides of the subject. If he is interested only in using mathematics as a tool, an understanding of the nature of mathematics will enable him to use it more effectively. If he is interested in mathematics for its own sake, the knowledge of areas of application of mathematics may furnish him with an insight into some new mathematical principle, as it has on many occasions in the history of the development of mathematics. For these reasons we now present a few of the applications of mathematics found in most mathematics textbooks for the elementary grades.

## PERCENTAGE

In mathematics, the word *per* may be translated as *divided by*. For example, "48 cents per dozen" means (48 ÷ 12) cents, or 4 cents each. Or, we speak of traveling 400 miles in 8 hours as 400 ÷ 8, or 50 miles per hour. This interpretation, coupled with that of the word *of* to mean *multiplied by* or *times*, makes the translation of percentage problems from words to symbols free from uncertainty.

A typical problem of this type is, "What per cent of 4800 is 1104?" The word *cent* means 100 (referring to money, century, etc.), so if we leave a blank or put a letter in place of the word *what* to represent the unknown number and translate carefully, we have

$$m \div 100 \times 4800 = 1104$$

or

$$\frac{m}{100} \times 4800 = 1104$$

which we can simplify to

$$m \times 48 = 1104$$

From the definition of division, we have

$$m = 1104 \div 48$$
$$= 23$$

Thus "23 per cent of 4800 is 1104," which can be verified by calculating 23/100 × 4800 (or 0.23 × 4800) to obtain 1104.

The superiority of this approach to the solution of percentage problems lies in the fact that every type of problem can easily be translated into the form

$$\frac{m}{100} \times n = p$$

where two of the three numbers $m$, $n$ and $p$ are known. The third number can then be found using basic knowledge of fractions, multiplication and division.

EXAMPLE: Suppose we are given the information that "2079.4 is 37% of some number" and are asked to find the number. Here the standard symbol, %, is used for the words *per cent*. With this interpretation we can write the sentence in algebraic form as

$$2079.4 = 37 \div 100 \times n$$

or

$$0.37 \times n = 2079.4$$

By the definition of division,

$$n = 2079.4 \div 0.37$$
$$= 5620$$

# APPROXIMATION

We tried to make it clear in our discussion of geometry that geometric figures and space are products of the imagination rather than concrete objects of physical space, although many ideas are conceived from striking properties of physical objects. The Greek philosopher, Plato, observed that no drawing or physical object could display the perfection required by geometry, so that the only perfect circles, for example, are those which occupy the mind. In order to make the wheel, however, man had to reproduce something similar to a circle. We might say that the outer edge of a wheel is an *approximation* of a circle. Whenever a carpenter measures the length of a board in order to cut another of the same length, he is making an approximation. If two segments $\overline{ab}$ and $\overline{cd}$ are not congruent, then one of them, say $\overline{ab}$, is longer than the other; that is

$$|ab| > |cd|$$

which means there is a point $p$ for which

$$a(p)b \quad \text{and} \quad |ap| = |cd|$$

so that $|ab| \neq |cd|$ no matter how near to $b$ point $p$ may be. Similarly, in terms of number alone, it is true that

$$\frac{1}{3} \approx 0.333$$

(which is read, "One-third is approximately equal to three hundred thirty three thousandths"), while the statement

$$\frac{1}{3} = 0.333$$

is *false* even though the difference is less than 0.001.

One of the uses of approximation occurs in calculations with non-terminating decimals in applied situations where precision is not necessary. When a person wants to know the diagonal measure across the floor of a square room 20 feet long, he might calculate theoretically

$$20 \times \sqrt{2}$$

whereas the product

$$20 \times 1.414 = 28.28$$

would be sufficiently accurate for his purpose. For purposes of comparing very large numbers such as 706,276 it is often convenient to use an approximation such as 706,000 which may not seem to be very close to the number when compared to the unit 1, but the comparison in such cases is often more realistically made with the number 706,276 itself. What we shall discuss here is the meaning of the language commonly associated with approximation and the effect of approximations upon computations of various kinds.

When an approximation is written with a decimal point, it is understood to differ from the exact number by *less than one-half of the place value of the right-most digit*. For example, if

$$470.63$$

is an approximation for a number $n$ then the positive difference of $n$ and 470.63 is less than

$$\frac{1}{2} \times \frac{1}{100} = 0.5 \times 0.01 = 0.005$$

Not knowing whether $n$ is greater or less than 470.63, we can only know that

$$470.63 - 0.005 < n < 470.63 + 0.005$$

or

$$470.625 < n < 470.635$$

A large number approximation such as 300,000 which does not employ the decimal point leaves us uninformed about its accuracy. Partly for this reason, a standard form called *scientific notation* has been devised in which the numeral is written with the left-most non-zero digit in units place so that the convention previously described may be used. Of course, this numeral must be completed by a factor which is the appropriate power of ten. Thus the numeral

$$3.000 \times 10^5 = 3 \times 100,000 = 300,000$$

represents the same approximation suggested previously, but indicates that the number which it approximates is between

$$(3.000 - 0.0005) \times 10^5 \quad \text{and} \quad (3.000 + 0.0005) \times 10^5$$

or

$$299,950 < n < 3000,050$$

The reason for the term *scientific notation* becomes apparent when we consider the very large numbers with which an astronomer, for example, must deal. At the other extreme, the nuclear physicist works with numbers as small as

$$0.00000000000105$$

which is quite unwieldy in this form. While

$$1.05 \times \frac{1}{10^{12}}$$

is somewhat more compact, the usual form is

$$1.05 \times 10^{-12}$$

*Special Techniques and Applications* 189

where the negative exponent is a little more convenient to handle in making computations.

*Definition 10.1.* A decimal numeral $m$ is an *approximation* of a number $p$ *accurate* to place $10^k$ iff $m - \frac{1}{2} \times 10^k < p < m + \frac{1}{2} \times 10^k$.

## Exercise 10.1

*1. Write a decimal numeral for each of the following:
  (a) $6\frac{3}{8}\%$   (b) $213.7\%$   (c) $5\frac{2}{3}\%$   (d) $0.0035\%$

*2. Write a percent numeral for the following:
  (a) $5{,}160$   (b) $0.07 \div 3$   (c) $\dfrac{3}{1000}$   (d) $\dfrac{6357}{800}$

3. Write common fractions to represent the numbers of Problem 1.

*4. Find the solution sets of the following equations:
  (a) $\dfrac{n}{100} \times 376 = 17.672$   (c) $333\% \times n = 135.531$
  (b) $17.6\% \times 857 = n$

5. (a) If $22\frac{1}{2}\%$ of the students in a class of 40 failed a test, how many students failed the test?
   (b) What per cent of a $600 budget goes for groceries if the amount spent for groceries is $150?

*6. The accuracy of the approximation 470.63 used as an example in the text was $\pm 0.005$. Give the accuracy of each of the following approximations, if possible:
   (a) 4.6   (b) 10.0063   (c) 0.0010   (d) 3,270,000

*7. Write the usual decimal numeral for each of the following:
   (a) $3.007 \times 10^3$   (b) $1.940 \times 10^2$   (c) $9.140 \times 10^{-3}$

*8. Write the scientific notation numeral for:
   (a) 301,200   (b) $17.06 \times 10$   (c) 0.00450

*9. Write the 4-digit approximation for each of the following, if possible:
   (a) $1.7321\cdots$   (d) $0.04163\cdots$
   (b) $12.3456\cdots$   (e) $107.5$
   (c) $35.4999\cdots$

10. For the numerals $7.03416\cdots$ and $7.03524\cdots$ the 4-place approximations 7.0342 and 7.0352 allow a 4-place rational number 7.0343 between them, whereas the 3-place approximations 7.034 and 7.035 would not allow a 3-place rational between. For the following pairs, determine similar pairs of approximations and intervening rationals.
   (a) $10.49615\cdots$; $10.49273\cdots$
   (b) $0.07200362\cdots$; $0.07199518\cdots$
   (c) $1.76543\cdots$; $1.73456\cdots$

*Elementary Mathematical Structure*

## CALCULATIONS WITH APPROXIMATIONS

Knowing the accuracy of two approximations, we would like to be able to predict the accuracy of approximations of sums, products and quotients obtained from them. A few problems of addition will furnish examples from which general properties can be deduced.

EXAMPLE: Suppose we wish to calculate the sum $35.6 + 17.34 + 5.782$ where each numeral is an approximation of the number $m$, $n$ and $p$ respectively. Then

$$35.55 < m < 35.65$$

$$17.335 < n < 17.345$$

and

$$5.7815 < p < 5.7825$$

which by addition indicates that

$$58.6665 < m + n + p < 58.7775$$

Now the only three-digit numeral which falls in the permissible range is

$$58.7$$

but this number is not an approximation for $m + n + p$ because

$$58.65 < m + n + p < 58.75$$

allows too much latitude on the left and not enough on the right. For example, it would be impossible for $m$, $n$ and $p$ to be small enough that their sum could be 58.651, according to the limits prescribed earlier; however, this number is allowed by the assumption that 58.7 is an approximation of $m + n + p$. On the other hand, it is possible for the exact numbers to be

$$m = 35.64$$

$$n = 17.344$$

and

$$p = 5.7824$$

giving

$$m + n + p = 58.7664$$

which is greater than the upper limit 58.75, established under the assumption that 58.7 is an approximation of the sum. In seeking the approximation of the sum

$$35.6 + 17.34 + 5.782 = 58.722$$

which is also an approximation of $m + n + p$, we must accept one less decimal place accuracy than the least accurate of the approximations being added. Thus, although 35.6 is accurate to the nearest tenth, when added with other

## Special Techniques and Applications

approximations it produces a sum which is accurate only to the nearest *unit* according to our definition of approximation. Then the most accurate approximation of the sum

$$35.6 + 17.34 + 5.782 = 58.722$$

which is also an approximation of the sum $m + n + p$ is 59. This procedure of finding an approximation of a numeral to a desired place value is called *rounding off the number to the nearest one* (or *tenth, hundredth*, etc. depending on the place value of the last digit).

From the example, we might wonder if there might be some simpler way of finding the limitations on $m + n + p$. Expressing the sum of the left hand members of the inequalities in terms of the original approximations, we have

$$(35.6 - 0.05) + (17.34 - 0.005) + (5.782 - 0.0005)$$
$$= (35.6 + 17.34 + 5.782) - (0.05 + 0.005 + 0.0005)$$
$$= 58.722 - 0.0555$$
$$= 58.6665$$

Thus the lower limit of the sum $m + n + p$ is less than the sum of approximations by the sum of the individual limits of error,

$$0.05 + 0.005 + 0.0005 = 0.0555$$

Similarly, the upper limit

$$58.7775 = 58.722 + 0.0555$$

is greater than the sum of approximations by the same amount, 0.0555. Then the difference of the upper and lower limits is

$$58.7775 - 58.6665 = (58.722 + 0.0555) - (58.722 - 0.0555)$$
$$= (58.722 - 58.722) + (0.0555 + 0.0555)$$
$$= 2 \times 0.0555$$

or *twice the sum of the limits of error*. Since this figure,

$$2 \times 0.0555 = 0.1110$$

is *greater than one tenth*, there is no approximation for $m + n + p$ which is accurate to the nearest tenth.

From the foregoing analysis, it seems that the factors which determine the decimal place accuracy of a sum of approximations are (1) the decimal place accuracy of the least accurate of the approximations and, perhaps, (2) the number of approximations being added. To examine the effect of (2), consider the sum of a large number of approximations each having the same place value accuracy. Since the effects (1) and (2) do not rely upon any peculiar feature of the approximations other than their place value accuracy, we will simplify the procedure by using the same approximation over and over again. Furthermore, we shall simply designate this approximation as $m$ and specify that it approximates a number $p$ with accuracy to three decimal places. From this information, we know that

$$m - .0005 < p < m + .0005.$$

Now we wish to know the place value accuracy of $m + m = 2m, m + m + m = 3m, 4m, \cdots, 10m,$ etc. Successive additions give

$$2m - 0.0010 < 2p < 2m + 0.0010$$
$$3m - 0.0015 < 3p < 3m + 0.0015$$
$$4m - 0.0020 < 4p < 4m + 0.0020$$
$$\cdots$$
$$10m - 0.005 \ < 10p < 10m + 0.005$$
$$11m - 0.0055 < 11p < 11m + 0.0055$$
$$\cdots$$
$$100m - 0.0500 < 100p < 100m + 0.0500$$
$$101m - 0.0505 < 101p < 101m + 0.0505$$
$$\cdots$$

These computations show that sums of from *two* through *ten* approximations (each accurate to three decimal places) have two-decimal place accuracy, since the differences of upper and lower limits range from

$$2 \times 0.0010 = 0.002 \quad \text{for} \quad 2m$$

to

$$2 \times 0.005 = 0.01 \quad \text{for} \quad 10m$$

none of which is greater than 0.01. Similarly,

$$2 \times 0.0055 = 0.0110 \quad \text{for} \quad 11m$$

and

$$2 \times 0.0500 = 0.1000 \quad \text{for} \quad 100m$$

indicate that from 11 to 100, such approximations may be added with one-decimal place accuracy guaranteed. These observations lead to a generalization which we can state as a theorem for reference when we discuss multiplication of approximations.

## Theorem 10.1

If $k$ approximations are added where $10^n < k \leq 10^{n+1}$ and the least accurate approximation is accurate to place value $10^m$ then the sum is accurate to place value $10^{m+n+1}$.

EXAMPLE: If we apply this theorem to our first example where $m \approx 35.6$, $n \approx 17.34$, and $p \approx 5.782$ ($\approx$ being a symbol for *is approximated by*) we find that $k = 3$ so that $n = 0$, since

$$10^0 = 1 < 3 < 10^{0+1} = 10^1 = 10$$

The least accurate approximation is 35.6, which is accurate to the $10^{-1} = 1/10$ place. By Theorem 10.1, the sum

$$35.6 + 17.34 + 5.782 = 58.722$$

approximates $m + n + p$ accurately to place value

$$10^{-1+0+1} = 10^0 = 1$$

or to ones' place. That is, rounding off 58.722 to units place gives an approximation to $m + n + p$, so

$$58.722 \approx 59 \approx m + n + p$$

as we have already seen.

EXAMPLE: Suppose $m \approx 3.08$ and we wish to predict an approximation of

$$m + m + \cdots + m = 957 \times m$$

which is guaranteed to be accurate. In this case, the number of approximations being added is 957, or $k = 957$. Then $n = 2$ since

$$100 = 10^2 < 957 < 10^3 = 1000$$

Also, the place value accuracy is $10^{-2}$ for each approximation, so $m = {}^-2$. According to the theorem we should be safe in rounding off the sum to the place

$$10^{-2+2+1} = 10^1$$

Now $3.08 + 3.08 + \cdots + 3.08 = 957 \times 3.08 = 2947.56$, so that $m + m + m + \cdots + m = 957 \times m \approx 2.95 \times 10^3$. (Remember that 2,950 does not clearly show that accuracy is to tens' place.)

## COMMON PRACTICES IN USING APPROXIMATIONS

In examining the problem of adding approximations, we have limited our predictions to the accuracy defined for the approximations being added. Part of our purpose in this self-imposed limitation was to stress consistency and to minimize complications. In making application of approximation theory to problems of measurement in laboratory situations, as in physics or statistics, several modifications are used. We will illustrate some of these methods and, where it is feasible, indicate the nature of the theory which supports them.

For our earlier example,
$$35.6 + 17.34 + 5.782 = 58.722$$
although we cannot use the three-digit numeral 58.7 as an approximation of the sum
$$m + n + p$$
in the sense of Definition 10.1, the information that
$$58.6665 < m + n + p < 58.7775$$
allows us to exhibit greater precision than that shown by the two-digit approximation 59. Since
$$58.7 - 58.6665 = 0.0335$$
and
$$58.7775 - 58.7 = 0.0775$$
we can say that 58.7 approximates $m + n + p$ within 0.1, or that
$$m + n + p = 58.7 \pm 0.1$$
(read 58.7 *plus or minus* 0.1). Similarly, from our second example, where
$$m - 0.0005 < p < m + 0.0005$$
the information that
$$6m - 0.003 < 6p < 6m + 0.003$$
can be expressed by the statement
$$6p = 6m \pm 0.003$$

A second principle which has affected the theory and practice of using approximations is demonstrated by comparing the difference of the approximations of two real numbers. For example, in comparing the numbers
$$3.141592653\cdots \quad \text{and} \quad 3.1412857141\cdots$$
we may compare the following differences:

$$
\begin{array}{lll}
3 & -\ 3 & =\ 0 \\
3.1 & -\ 3.1 & =\ 0.0 \\
3.14 & -\ 3.14 & =\ 0.00 \\
3.142 & -\ 3.141 & =\ 0.001 \\
3.1416 & -\ 3.1413 & =\ 0.0003 \\
3.14159 & -\ 3.14129 & =\ 0.00030 \\
3.141593 & -\ 3.141286 & =\ 0.000307 \\
3.1415927 & -\ 3.1412857 & =\ 0.0003070 \\
\end{array}
$$

*Special Techniques and Applications*

Since the first few digits are the same for corresponding place values, the shorter approximations are useless in such a comparison. Notice that this phenomenon has nothing to do with the location of the decimal point (provided it occurs in corresponding positions in both numerals, of course), but rather with the *number of digits* used in the approximations when written in standard (scientific notation) form. Such digits are said to be *significant*.

*Definition 10.2.* A digit of a terminating decimal numeral is said to be *significant* if it occurs in the standard (scientific notation) numeral for the number.

EXAMPLE: Since

$$0.002703 = 2.703 \times 10^{-3}$$

the digits 2, 7, 0 and 3 are significant. In the four place approximation of the number

$$0.101973\cdots \approx 0.1020$$

all four digits 1, 0, 2 and 0 are significant, while only the first three digits of the approximation

$$70,300 \approx 70,284$$

are significant. Such expressions as "accurate to five digits" or "correct to three figures" refer to the number of significant digits which are accurate in accordance with Definition 10.1.

## SUBTRACTION

To summarize the case for subtracting approximations, let us consider $m > n$ as approximations of $p$ and $q$ respectively, with $e$ and $\epsilon$ as corresponding limits of error. Then

$$m - e < p < m + e$$

and

$$n - \epsilon < q < n + \epsilon$$

Since $m + e$ is the upper limit on $p$ and $n - \epsilon$ the lower limit on $q$, the upper limit on $p - q$ is

$$(m + e) - (n - \epsilon) = (m - n) + (e + \epsilon)$$

or

$$p - q < (m - n) + (e + \epsilon)$$

Similarly, by the properties of $<$,

$$(m - e) - (n + \epsilon) = (m - n) - (e + \epsilon)$$

and
$$(m - n) - (e + \epsilon) < p - q$$
so that
$$(m - n) - (e + \epsilon) < p - q < (m - n) + (e + \epsilon)$$

We can see that the limit of error for the difference of two approximations is the same as that for their sum. If we observe that the theory which we have developed does not require approximations to be positive numbers, and recall that
$$x - y = x + {}^-y$$
for real numbers $x$ and $y$, we see that this result is a special case of Theorem 10.1.

## MULTIPLICATION

Since multiplication by a natural number can be effected by repeated addition, we might wonder if Theorem 10.1 will also take care of multiplication. Unfortunately, this case is not quite so simple. For analysis, again let
$$m - e < p < m + e$$
and
$$n - \epsilon < q < n + \epsilon$$
so that
$$(m - e)(n - \epsilon) < pq < (m + e)(n + \epsilon)$$

By using the properties of subtraction, negatives, and distribution, we have
$$(m - e)(n - \epsilon) = mn + e\epsilon - (m\epsilon + ne)$$
and
$$(m + e)(n + \epsilon) = mn + e\epsilon + (m\epsilon + ne)$$
so that the range from lower to upper limit is
$$2(m\epsilon + ne)$$

Since this number identifies the place value which establishes the accuracy of the approximations of $mn$ as approximations of $pq$, we see that both $m \times 2\epsilon$ and $n \times 2e$ contribute to this number.

EXAMPLE: In the product
$$436.23 \times 0.7062$$
let $m = 436.23$ and $n = 0.7062$, whence
$$2e = 0.01 \quad \text{and} \quad 2\epsilon = 0.0001$$

*Special Techniques and Applications* **197**

Then
$$m \times 2\epsilon + n \times 2e = 0.043623 + 0.007062$$
$$= 0.050685$$

which indicates that the one-place approximation of
$$436.23 \times 0.7062 = 308.065626$$
that is, 308.1 is an approximation of $pq$. More precisely, we know that
$$pq = 308.07 \pm 0.03$$
since $\frac{1}{2} \times .050685 < 0.03$. These results are verified by the following calculations.
$$436.225 < p < 436.235$$
$$0.70615 < q < 0.70625$$
$$308.04028375 < pq < 308.09096875$$
which compares with the previous result
$$pq = 308.07 \pm 0.03$$
by the inequalities
$$308.07 - 0.03 < 308.04028375 < 308.09096875 < 308.07 + 0.03$$

EXAMPLE: If $p \approx 4{,}036$ and $q \approx 91.8$, we may establish limits on $pq$ from the basic relations
$$4{,}035.5 < p < 4{,}036.5$$
and
$$91.75 < q < 91.85$$

Then multiplication of each pair produces
$$370{,}257.125 < pq < 370{,}752.525$$
so that the difference between limits is
$$370{,}752.525 - 370{,}257.125 = 495.400$$

Adding half of this difference to the smaller number produces an average of about
$$370{,}500$$
which is rounded off to the fourth digit, since only the first three digits of the two limits agree. Incidentally, the product
$$4{,}036 \times 91.8 = 370{,}504.8$$
also rounds off to the same number. Then, since
$$\frac{1}{2} \times 495.400 < 300$$

the accuracy of the approximation for $pq$ is given by

$$pq = 370{,}500 \pm 300$$

As an approximation in the sense of Definition 10.1, however, the best that we can give is

$$pq \approx 3.7 \times 10^5$$

since one limit rounds to 370,000 and the other rounds to 371,000.

From the formal analysis, if $m = 4{,}036$ and $n = 91.8$ then $2e = 1$ and $2\epsilon = 0.1$ so that $2(m\epsilon + ne) = 403.6 + 91.8 = 495.4$, as we have just seen. Then the product

$$4{,}036 \times 91.8 = 370{,}504.8$$

rounds off to either 370,000 or 371,000, so that two-digit accuracy is all that is possible.

Let us now generalize by allowing $m$ and $n$ to be two positive approximations of $p$ and $q$ respectively, with $e$ and $\epsilon$ the respective positive limits of error. If

$$m \times 2\epsilon > n \times 2e$$

and if $10^k$ is the place value of the first significant digit of $m \times 2\epsilon$, then

$$1 \times 10^k \leq m \times 2\epsilon < 1 \times 10^{k+1}$$

and

$$0 < n + 2e < 1 \times 10^{k+1}$$

Then, by addition

$$1 \times 10^k < m \times 2\epsilon + n \times 2e < 2 \times 10^{k+1}$$

so that $mn$ rounded off at place $10^k$ is not an approximation for $pq$, but when rounded off at place $10^{k+1}$ the result differs from $pq$ by less than $1 \times 10^{k+1}$.

## Theorem 10.2

If $m$ and $n$ are approximations with $p = m \pm e$ and $q = n \pm \epsilon$, and if $10^k$ is the place value of the first significant digit of the larger of $2m\epsilon$ and $2ne$, then $mn$ rounded off at place $10^{k+1}$ differs from $pq$ by less than $1 \times 10^{k+1}$.

EXAMPLE: From the product

$$436.23 \times 0.7062$$

of our first example in multiplication, we calculated

$$2m\epsilon = 0.043623$$

and

$$2ne = 0.007062$$

*Special Techniques and Applications* **199**

The larger number has $0.01 = 10^{-2}$ as the place value of its first significant digit, 4. According to Theorem 10.2, the product

$$mn = 436.23 \times 0.7062 = 308.065626$$

when rounded off to the $10^{-2+1} = 10^{-1}$ place, that is, 308.1 differs from $pq$ by less than $10^{-2+1} = 10^{-1} = 0.1$. In other words,

$$pq = 308.1 \pm 0.1$$

or

$$308.0 < pq < 308.2$$

which is certainly valid according to our previous findings, but not as precise as the more laborious methods.

## RULES OF COMPUTATION

There are a number of rules for making computations with approximations which are designed to eliminate the use of digits having no effect on the accuracy of the result. For some, the accuracy of the result may be predetermined, but for others, the accuracy varies with the particular problem and can be ascertained only by the principles which we have used in our analysis. One rule for adding a number of approximations asks that all be rounded off to the same accuracy as the least accurate. We show two examples to indicate how the results may vary.

EXAMPLE:

|     | (1) | 40.3 | 40.3 | (2) | 40.3 | 40.3 |
|-----|-----|------|------|-----|------|------|
|     |     | 17.235 | 17.2 |     | 17.245 | 17.2 |
|     |     | 91.66 | 91.7 |     | 91.634 | 91.6 |
|     |     | 103.0101 | 103.0 |     | 103.049 | 103.0 |
|     |     | 252.2051 | 252.2 |     | 252.228 | 252.1 |

In (1) the excesses and shortages almost cancel each other, whereas in (2) the rounding off always produces shortage. If the columns were long and if the preponderance of entries had three- and four-place accuracy, the error would be greatly magnified.

If our object had been to develop skill in working with approximations, we would have selected some of the better rules, explained and illustrated their use, and provided exercises for practice. In this development, however, our aim has been to help you develop an understanding of approximation, and to learn to use it for determining the accuracy of a computation. The following exercises are provided to promote this aim.

*Exercise 10.2*

1. Establish the most accurate limits possible for the following sums of approximations:
   (a) $43.76 + 19.034 + 27.369$
   (b) $1.004 \times 10^5 + 8.38 \times 10^4 + 6.106 \times 10^4$
   (c) $3.45 + 2.17 + 4.63 + 2.38 + 3.09$

2. Establish the most accurate limits possible for the following operations with approximations:
   (a) $1.7321 - 1.4142$     (c) $0.013 \times 101.7$
   (b) $1.0001 - 0.9999$     (d) $99.9 \times 9.99$

3. If $0 < a < b$ and $0 < c < d$ then of all the quotients $a \div c$, $a \div d$, $b \div c$, and $b \div d$, $a \div d$ is the least and $b \div c$ is the greatest. Use this principle to establish limits for the following quotients of approximations, showing first four places of the decimal numeral for each quotient used.
   (a) $41.07 \div 0.73$     (b) $9.689 \div 1.1$     (c) $1.01 \div 99.04$

4. By observing the limits established in Problem 1, write the most accurate (Definition 10.1) approximation possible for each sum.

5. Apply the directions of Problem 4 to the results which you obtained in Problem 2.

6. Apply the directions of Problem 4 to your results in Problem 3.

7. Apply Theorem 10.1 to find approximations for the sums of Problem 1.

8. Use $m \times 2\epsilon + n \times 2e$ to find best approximations in parts (c) and (d) of Problem 2.

9. Express the sums of Problem 1 to one more decimal place than their approximations of Problem 4, showing the limit of error by using the form $s \pm e$.

10. Apply the directions of Problem 9 to the results of Problem 2 as compared to those of Problem 5.

## PROBABILITY AND CHANCE

Most of us are well aware of the concept of *chance* as it relates to our daily lives. We say that we are taking a chance whenever we drive our automobiles in traffic because we hear that a certain percentage of motorists were involved in traffic accidents last year. Our chances of becoming victims of cancer, heart disease, tuberculosis, etc., are repeatedly brought to our attention by various interested organizations. In fact, the idea of *probability* is so commonplace in the maze of *statistical* situations of our complex society that we seldom think of it as the basis for gambling or games of chance. Yet, interest in such games initiated correspondence between two French mathematicians, Pierre de Fermat (1608–1665) and Blaise Pascal (1623–1662), which resulted in the development of the first recorded mathematical treatment of probability. Another French mathematician and scientist, Pierre Simon de

Laplace (1749–1827), reorganized and refined the theory in a treatise regarded by some of his contemporaries as more complex than any mathematical theory previously produced.

We will confine our attention to the most fundamental concepts, to see how mathematics can be used to define certain useful quantities. It will be easier to illustrate some of these basic ideas than to give abstract definitions, so we shall give examples and introduce some of the terminology of probability in the process. To keep examples as simple as possible, we will refer to such gambling devices as dice and spinners. Most children will be acquainted with these devices since they are widely used in children's games.

EXAMPLE 1: Perhaps the simplest example of probability occurs in the act of tossing a coin to determine whether "heads" or "tails" appears. From the practical viewpoint, such an act can have other outcomes, such as losing the coin or having it come to rest on edge. In the interest of theory, however, such unwanted outcomes can be eliminated by ignoring those results and performing the act again. The outcomes which we wish to regard as official are usually called *events*.

EXAMPLE 2: If a die (plural, dice) is tossed and comes to rest on a smooth flat surface, we would recognize six outcomes, depending upon the number of spots showing on the upper face.

EXAMPLE 3: If a spinner is spun and comes to rest pointing to one of the colors of the circular region as pictured, we would recognize three outcomes.

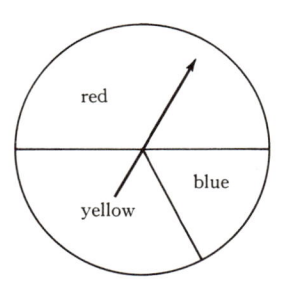

EXAMPLE 4: If a card is dealt from a bridge deck, we may wish to recognize more than one kind of event. We might be interested in the suit (spade, heart, club, diamond), the color (red, black), or the face value (2, 3, ···, king, ace).

## EQUALLY LIKELY EVENTS

Accepting the intuitive notion conveyed by the expression "equally likely," we would be prone to say that the two events of Example 1 are equally likely.

If we were to toss a particular coin one hundred times and observe heads appear seventy-six times, we would be very reluctant to call the events equally likely. Such a decision would be based upon experience. In looking at Example 3, however, we would expect the spinner to stop on red most often, on yellow less often and on blue least often, even before experimenting with such a device. If the circular region had been colored as shown, the events red, yellow

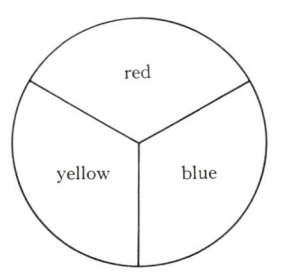

and blue would seem equally likely because the three regions appear to be congruent. If the regions of Example 3 are subdivided to appear as shown, then yellow should result about twice as often as blue, and red should appear three times as often as blue.

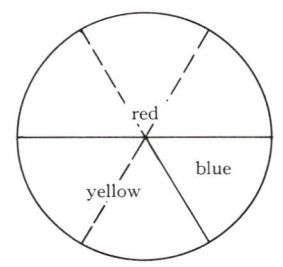

From these considerations it seems reasonable to view probability as a kind of *measure*, such as the *frequency* of occurrence or the *area* of a region, etc. However, such a measure of frequency or area has little significance unless it is compared to its potential. For example, the imperfection or *bias* of the coin which came up heads 76 times is unknown without the information that it was tossed 100 times. Thus we say that the probability that the biased coin will turn up heads when tossed is about "76 out of 100" or, more formally, 76% or 76/100; that is, we form a ratio of the measure related to the occurrence of the event and the measure of its potential. Applying this principle to our examples, we have 1/2 as the probability of a head appearing from a toss of the unbiased coin of Example 1; 1/6 is the probability that the die of Example 2 will turn up its five spot side, since there are six sides and only one of them has five spots on it; and the probabilities that the spinner of Example 3 will point to blue, yellow, or red are 1/6, 2/6 = 1/3 and 3/6 = 1/2, respectively.

## MUTUALLY EXCLUSIVE EVENTS

One of the initial difficulties which students have in relating probability to a particular event is deciding what other events must also be considered in order to compute the probability of the event occurring. In the first three of our four examples, there is little likelihood of confusion. In Example 1, we are interested in only two events—heads or tails. The state of the weather, the kind of coin being used, or any other consideration, has no bearing on the outcome; only these two events are relevant. Notice, however, that the two events are related to each other in that both cannot occur in one toss. We say that they are *mutually exclusive* events; that is, the occurrence of one automatically gives the information that the other did not occur. The six events of Example 2 and the three events of Example 3 also have this property.

The case of Example 4 is a little more complex than the others. Besides the *number* of events associated with dealing a card from a bridge deck, there are also different *kinds* of events. Suppose, for example that we select as an event the drawing of a card with red pips. What other events should be associated with this one in order to calculate its probability? Since these events must not occur when a red card is dealt, we need to examine the cards which remain after all the red cards are removed from the deck. What common characteristics do these cards have? Should we also bring in some other objects, such as a pinochle deck? The answer to the latter question is negative, since we know that dealing a card from the bridge deck will in no way involve the pinochle deck. Thus we confine our attention to the non-red bridge cards, and observe that they are all black cards. In other words, when a card is dealt, it will either be red or black exclusively; thus, dealing a red card and dealing a black card are two mutually exclusive events. Upon counting, we find that the complete deck contains 52 cards, of which 26 are red, so that the probability of dealing a red card (unless the deck is stacked!) is 26/52 or 1/2. Similarly, the probability of dealing a black card is $26/52 = 1/2$.

Suppose that in examining the non-red cards of Example 4, we did not observe that all the other cards were black but instead noticed that some of them were spades and the rest were clubs, whereas none of the red cards were either spades or clubs. In other words, for the sets

$D$ is a complete deck of bridge cards

$R = \{r \mid r \in D, r \text{ is red}\}$

$C = \{c \mid c \in D, c \text{ is a club}\}$

$S = \{s \mid s \in D, s \text{ is a spade}\}$

we have the relations

$$D = R \cup C \cup S$$

$$R \cap C = \emptyset, \quad R \cap S = \emptyset, \quad \text{and} \quad C \cap S = \emptyset$$

By our informal definition, if $E$ denotes an event and $P[E]$ denotes its probability of occurrence,

$$P[R] = \frac{n(R)}{n(D)} = \frac{26}{52} = \frac{1}{2}$$

$$P[C] = \frac{n(C)}{n(D)} = \frac{13}{52} = \frac{1}{4}$$

$$P[S] = \frac{n(S)}{n(D)} = \frac{13}{52} = \frac{1}{4}$$

It should be observed that even though the set of cards in a bridge deck can be collected into disjoint or mutually exclusive subsets, this fact alone will not insure the preceding probabilities for the indicated outcomes in dealing the first card. The deck needs to be thoroughly shuffled and perhaps cut before dealing, as every card player knows. This procedure is to insure that each of the 52 cards has an equal chance of being the top card, or that the selection of the card will be *random*. It is this factor in the method of selection which allows the ratios of **1:1:2** of clubs, spades and red cards to be effective.

## SUMMARY

Let us pause briefly to summarize the properties involved in defining a probability for an event. First of all, there must be a universal set $U$ which consists of all elements $x$ for which the sentence $p(x) = x$ *implies the event occurs* is a statement (in the true-false logic sense). For example, the six congruent arcs $\widehat{ab}, \widehat{bc}, \widehat{cd}, \widehat{de}, \widehat{ef}, \widehat{fa}$ of the circle of Example 3 could serve as elements of a universal set for the event of the spinner's coming to rest, since it must point to some point on the circle which is in turn in one of the six arcs.

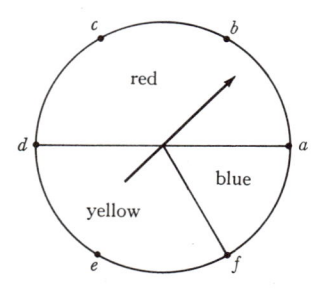

In case it points at one of the six points $a, b, c, d, e, f$ we may arbitrarily assign point $a$ to $\widehat{ab}$, point $b$ to $\widehat{bc}$, and so on around the circle. This condition provides for two disjoint subsets of $U$, namely

$$T = \{x \mid x \in U, p(x) \text{ is true}\}$$

and
$$F = U - T = \{x \mid x \in U, p(x) \text{ is false}\}$$

In the second place, each event represented by the elements of $U$ must have an equal chance of happening. That is, an element is selected from $U$ at random. In the example, we construct the spinner so that it is well balanced and turns with equal ease through each arc of the circle.

Finally, the sets $T$ and $U$ must have a measure $m(T)$ and $m(U)$. In the case of finite sets, the measure is the cardinal number of the set.

When all three conditions are satisfied, the *probability of an event E* is defined by the ratio

$$P[E] = \frac{m(T)}{m(U)}$$

EXAMPLE: In the example of the spinner, if we let
$$Y = \{\widehat{de}, \widehat{ef}\}$$
then for any point $p$ of the circle where our spinner stops we have
$$P[p \in Y] = \frac{m(Y)}{m(U)} = \frac{n(Y)}{n(U)} = \frac{2}{6} = \frac{1}{3}$$

EXAMPLE: In order to give some meaning to the use of *measure* as a more general concept, consider an example similar to the preceding, except that the arcs $\widehat{df}, \widehat{fa}$ and the semi-circle $\widehat{ad} = \widehat{ab} \cup \widehat{bc} \cup \widehat{cd}$ are incommensurable; that is, there is no arc of the circle for which the lengths of $\widehat{df}, \widehat{fa}$ and $\widehat{ad}$ are all whole number multiples of its length. In such a case we recall a formula from geometry which gives the measure of the circumference of a circle as $2\pi r$ where $r$ is the measure of its radius. If the radius is taken as a unit, this measure becomes
$$2\pi = 2(3.1415\cdots) = 6.2831\cdots$$
From this viewpoint, $U$ is the set of points belonging to the circle, $Y$ is the set of points belonging to $\widehat{df}$, and the measure of $Y$ is the length of $\widehat{df}$ (which is some positive real number less than $2\pi$). To illustrate further, let us use the same $Y$ as before, whose length is one-third the circumference of the circle. That is,
$$m(Y) = \frac{1}{3} \times 2\pi$$
so that
$$P[p \in Y] = \frac{m(Y)}{m(U)} = \frac{\frac{1}{3} \times 2\pi}{2\pi} = \frac{1}{3}$$
as before.

*Exercise 10.3*

1. Toss a coin 100 times (in as random a fashion as you can) and compare the result with the ideal of 50 heads and 50 tails.
2. Toss two coins together 20 times and record the frequencies (number of occurrences) of (a) 2 heads, (b) 2 tails, and (c) one head and one tail.
3. Thinking of two *kinds* of coins in Problem 2 (in order to distinguish them), (a) show that there are 4 possible outcomes of a toss and (b) calculate the probability of each of the three events described in Problem 2. (c) What are the ideal frequencies for (a), (b) and (c) of Problem 2?
4. In the roll of a die, what is the probability of turning up a number larger than (a) 5? (b) 3? (c) 2? (d) 0? (e) 6?
5. If two dice of different colors (just to help identify them) are rolled simultaneously, (a) show that there are 36 different outcomes and (b) show that these outcomes correspond to the elements of $N_6 \times N_6$ where $N_6 = \{1, 2, 3, 4, 5, 6\}$.
6. In a toss of the dice of Problem 5, show that the probability that the sum of numbers turned up is (a) 2, (b) 3, (c) 6, (d) 7, (e) 11 should be 1/36, 1/18, 5/36, 1/6, and 1/18 respectively.
7. If the bottom of a square box is partitioned into 16 congruent cells as shown, and if a marble could be dropped so that it is equally likely to fall into each one of the cells, what is the probability that it will fall into a cell marked (a) ////? (b) \\\\? (c) unshaded?

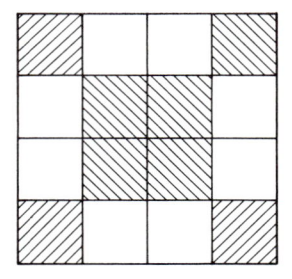

8. A dart board is composed of 32 congruent sectors separated by raised wire boundaries. 20 sectors are yellow, 8 are black and 4 are red. A dart thrown at random strikes the board. What is the probability that the sector it is in is (a) yellow? (b) black? (c) red?
9. Imagine two dice without spots such that on one die, three faces are red, two are blue and one is yellow. Let the other die have two red faces, one blue face, and three yellow faces. In one roll of the dice, what is the probability of obtaining (a) two red faces? (b) two blue faces? (c) two yellow faces? (d) one red and one blue? (e) one blue and one yellow?
10. Identify subsets $A$, $B$ of $N_6 = \{1, 2, 3, 4, 5, 6\}$ to represent the red faces of the two dice so that $A \times B$ represents the outcomes for two red faces. Make a similar model for the case of one red and one blue face.

# COMPOUND EVENTS

Sometimes events can be reclassified in terms of simpler events. For example, the roll of a pair of dice can be described by two throws of the same die. Such events will be called *compound* events; otherwise, an event will be called *simple*. Examples 1 through 4 all described simple events, while most of the problems of Exercise 10.3 were concerned with compound events. Although Definition 10.3 is sufficient to calculate all probabilities, it is often easier to calculate the probabilities of simpler related events and combine these results in some way to obtain the probability of the compound event. In this section, we will illustrate certain situations of this type, and state the basic definitions and theorems commonly used. There are two ways in which compound events are related to simpler events. These classifications are directly related to the idea that a set may be in some cases a union of some of its subsets, and in some cases a cartesian product of sets. We will examine the latter case first.

EXAMPLE 5: If a die is rolled twice, what is the probability that a one or two will appear on the first throw and a four, five or six on the second?

EXAMPLE 6: Two cards are dealt in succession from a well-shuffled bridge deck. What is the probability that the first card is a face card and the second is not?

The events described by Examples 5 and 6 might be called *composite* events, because the occurrence of two or more simple events is required to constitute one event of this type. The events of Examples 5 and 6 differ in another respect. In Example 5, the conditions under which the die is thrown are assumed to be the same each time. In Example 6, however, the first card is dealt from a deck of 52 cards, while the second card is dealt from a deck of only 51 cards. Furthermore, the probability of drawing a non-face card as the second card depends upon the kind of card which was first dealt. We say that the two simple events of Example 5 are *independent* events, and the simple events of Example 6 are *dependent*.

All possible outcomes of the event $E$ of Example 5 can be represented by

$$A \times B = N_6 \times N_6$$

where

$$A = N_6 = \{1, 2, 3, 4, 5, 6\}$$

represents the six possible outcomes of the simple event of the first throw of the die and

$$B = N_6 = \{1, 2, 3, 4, 5, 6\}$$

represents the possible outcomes of the second throw. The composite event of the problem can be represented by the cartesian product of the subsets

$$C = \{1, 2\} \subset A$$

and
$$D = \{4, 5, 6\} \subset B$$
since the elements of
$$C \times D = \{(1, 4), (1, 5), (1, 6), (2, 4), (2, 5), (2, 6)\}$$
indicate those elements of $A \times B$ which describe the occurrence of either a 1 or 2 followed by the occurrence of either a 4, 5 or 6. Then by Definition 10.3
$$P[E] = \frac{n(C \times D)}{n(A \times B)} = \frac{6}{36} = \frac{1}{6}$$

The theory of cardinal numbers, however, shows the following relationship:
$$P[E] = \frac{n(C \times D)}{n(A \times B)} = \frac{n(C) \cdot n(D)}{n(A) \cdot n(B)} = \frac{n(C)}{n(A)} \cdot \frac{n(D)}{n(B)} = P[E_1] \cdot P[E_2]$$
where $E_1$ and $E_2$, respectively, represent the component events of the occurrence of a 1 or 2 on the first throw of the die and the occurrence of a 4, 5 or 6 on the second throw. Note also that
$$P[E_1] \cdot P[E_2] = \frac{n(C)}{n(A)} \cdot \frac{n(D)}{n(B)} = \frac{2}{6} \cdot \frac{3}{6} = \frac{1}{3} \cdot \frac{1}{2} = \frac{1}{6}$$

In searching for a universal set for the event $E$ of Example 6, once again it is simpler to think of a composite event. Let $A$ designate a universal set for the event $E_1$, dealing a face card first from a deck of bridge cards. Since such a deck has 52 cards, we have
$$n(A) = 52$$
Similarly, let $B$ designate a universal set for the second component, event $E_2$, dealing a non-face card as the second card off the deck, so that
$$n(B) = 51$$
If $C$ and $D$ represent the sets of favorable outcomes of $E_1$ and $E_2$ respectively, we have
$$n(C) = 16$$
and
$$n(D) = 36$$
Notice that we are not concerned with any case where a non-face card is drawn first, so *all* face cards are represented by $C$ and *all* non-face cards by $D$. Then
$$A \times B \quad \text{and} \quad C \times D$$
represent respectively a universal set for event $E$ and the set of favorable

*Special Techniques and Applications*          **209**

outcomes, so that

$$P[E] = \frac{n(C \times D)}{n(A \times B)} = \frac{\overset{4}{\cancel{16}} \cdot \overset{12}{\cancel{36}}}{\underset{13}{\cancel{52}} \cdot \underset{17}{\cancel{51}}} = \frac{48}{221}$$

Once again we observe that

$$P[E] = \frac{n(C \times D)}{n(A \times B)} = \frac{n(C) \cdot n(D)}{n(A) \cdot n(B)} = \frac{n(C)}{n(A)} \cdot \frac{n(D)}{n(B)} = P[E_1] \cdot P[E_2]$$

These examples illustrate principles which can be summarized by the following definition and theorem.

*Definition 10.4.* An event $E$ is *composite* if it consists of the occurrence of a set of two or more events called *component* events.

## Theorem 10.3

The probability of the occurrence of an ordered sequence of events is the product of the probabilities of the occurrence of the component events.

## CLASSIFIED EVENTS

We turn our attention now to the study of events whose outcomes can be grouped naturally into subsets. Suppose, for example, that two sets $A$ and $B$ are so related that

$$A \cup B$$

represents a universal set for an event $E$. In calculating $P[E]$ we would need to find

$$n(A \cup B)$$

which is given in general by

$$n(A \cup B) = n(A) + n(B) - n(A \cap B)$$

according to Problem 4 of Exercise 2.3. If the events $E_1$ and $E_2$ are *mutually exclusive*, then

$$A \cap B = \emptyset$$

so that

$$n(A \cup B) = n(A) + n(B)$$

EXAMPLE 7: What is the probability that the spinner of Example 3 will not stop on blue?

In this case, the event $E$ can occur either by event $E_1$, where the spinner stops on red, or by event $E_2$, the spinner stopping on yellow. If we let $R$ represent the semi-circle bordering the red sector and $Y$ the arc bounding the yellow sector, then the entire circle $C$ will serve as a universal set for $E$ and

$$R \cup Y$$

will represent the set of all favorable outcomes. By Definition 10.3,

$$P[E] = \frac{m(R \cup Y)}{m(C)} = \frac{\frac{5}{6} \cdot 2\pi r}{2\pi r} = \frac{5}{6}$$

where $r$ is the radius of the circle. Since measure of arc length has the same additive property as cardinal number, it is also true that

$$P[E] = \frac{m(R \cup Y)}{m(C)} = \frac{m(R) + m(Y)}{m(C)}$$

$$= \frac{m(R)}{m(C)} + \frac{m(Y)}{m(C)} = P[E_1] + P[E_2]$$

This result is also verified numerically by using

$$P[E_1] = \frac{1}{2} \quad \text{and} \quad P[E_2] = \frac{1}{3}$$

from Example 3, observing that

$$\frac{1}{2} + \frac{1}{3} = \frac{3}{6} + \frac{2}{6} = \frac{5}{6}$$

EXAMPLE 8: A number is selected at random from the set

$$N_9 = \{1, 2, 3, 4, 5, 6, 7, 8, 9\}.$$

What is the probability that it is either a prime or an odd number?

Here the set $N_9$ can serve as the universal set for $E$, and the set of favorable outcomes is

$$\{1, 2, 3, 5, 7, 9\}$$

If we let

$$A = \{2, 3, 5, 7\}$$

and

$$B = \{1, 3, 5, 7, 9\}$$

then $A$ and $B$ represent $E_1$ and $E_2$, the events of drawing a prime number and an odd number, respectively. Then by Definition 10.3,

$$P[E] = \frac{n(A \cup B)}{n(N_9)} = \frac{6}{9} = \frac{2}{3}$$

*Special Techniques and Applications* 211

It is also true, however, that

$$P[E] = \frac{n(A \cup B)}{n(N_9)} = \frac{n(A) + n(B) - n(A \cap B)}{n(N_9)}$$

$$= \frac{n(A)}{n(N_9)} + \frac{n(B)}{n(N_9)} - \frac{n(A \cap B)}{n(N_9)}$$

$$= P[E_1] + P[E_2] - P[E_1 E_2]$$

where $P[E_1 E_2]$ represents the probability that both $E_1$ and $E_2$ will occur simultaneously; that is, the number drawn will be an odd prime. To check the latter calculation, we note that

$$A \cap B = \{3, 5, 7\}$$

so that

$$P[E_1] + P[E_2] - P[E_1 E_2] = \frac{4}{9} + \frac{5}{9} - \frac{3}{9} = \frac{6}{9} = \frac{2}{3}$$

as predicted. The following theorem is often called the *addition theorem* of probability.

## Theorem 10.4

If $A$, $B$ are sets which represent the outcomes of events $E_1$, $E_2$, then the probability of the occurrence of either $E_1$ or $E_2$ is $P[E_1] + P[E_2] - P[E_1 E_2]$.

A final example will illustrate both of these principles, and will introduce a method of counting which is useful in many statistical situations. This situation involves an event repeated a number of times and with two outcomes, referred to as a *binomial* probability. Repeated tosses of a coin or die (or a single toss of several coins or dice) are examples of this type of probability.

EXAMPLE 9: If a die is tossed five times, what is the probability that numbers less than 3 will turn up exactly three times?

This problem is similar to that of Example 5. The principle difference is that no *order* is specified in this case for the occurrence of the low scores. If we view this event as a compound event, we may (1) calculate the probability of the occurrence of each favorable outcome, and (2) combine these probabilities according to Theorem 10.4. In order to perform the first step, we will select a particular favorable outcome; for instance, numbers less than 3 turn up the *first* three times while other numbers turn up on the last two throws. This is a *composite* event, so we must know the probabilities of its *component* events, and apply Theorem 10.3. One type of component event is the occurrence of a 1 or 2 on a single throw. By Definition 10.3 this probability is

$$\frac{2}{6} = \frac{1}{3}$$

which is the probability for each of the first three component events. The other type of component event is the occurrence of a 3, 4, 5 or 6 on a single throw of the die. The probability of this event is

$$\frac{4}{6} = \frac{2}{3}$$

and applies to the last two of the component events. By Theorem 10.3, the probability that a number less than three results in exactly the first three of five throws of a die is

$$\frac{1}{3} \cdot \frac{1}{3} \cdot \frac{1}{3} \cdot \frac{2}{3} \cdot \frac{2}{3} = \frac{4}{243}$$

If the same analysis is applied to the event of the occurrence of a number less than three on exactly three of five throws of a die, no matter which three, the probability is the same. The only difference will be the order in which the factors 1/3, 1/3, 1/3, 2/3 and 2/3 occur. Thus, to effect the second step, since these favorable outcomes each of probability 4/243 are mutually exclusive events, we need only add as many 4/243's as there are favorable outcomes. Although it is possible to list and count all favorable outcomes, it seems that there should be some method of calculating the number. For the moment, we turn our attention to this new topic.

## PERMUTATIONS AND COMBINATIONS

Although we will not list the favorable outcomes of Example 9, we analyze a systematic procedure for doing so in order to derive a formula for calculating their number. For convenience we will use the symbols $*$ and 0 in five spaces from left to right, to indicate the occurrence ($*$) and non-occurrence (0) of a number less than three on each of five throws of the die in order from first to fifth. Thus

$$* \; * \; 0 \; * \; 0$$

indicates that numbers smaller than three turned up on the first, second and fourth throws.

Suppose we ask how many favorable outcomes have $*$ in the first two places. Since the third $*$ must occur in either the third, fourth or fifth places exclusively, it is apparent that the answer is 3. However, the answer would be 3 no matter which two places are occupied by the other two $*$. Then we need to know how many favorable outcomes have $*$ in the first place. This

*Special Techniques and Applications*

means that the second ✶ can be in any place *except* the first, for example

$$✶ - ✶ - -$$

still leaving a choice of 3 places for the third ✶. Thus when one ✶ is on the first place, the second ✶ can be in any one of 4 places and, for each of these 4 choices, there are 3 places for the third asterisk. The number of favorable outcomes having ✶ in the first place, then, is

$$4 \cdot 3 = 12$$

However, no ✶ has to be in the first place, so we recognize that the first ✶ could be in any of the five places. In each case there are 12 arrangements of the second and third ✶, or a total of

$$5 \cdot 12 = 60$$

By this time you should be wondering how we can distinguish between the first, second and third ✶ and, of course, we cannot. Therefore, we have some duplication which we must adjust. For example, how many times have we counted the arrangement

$$✶ \;✶\; 0 \;✶\; 0$$

The answer should be, "As many times as the number of ways that the numbers 1, 2 and 3 can be assigned to the three ✶'s." Notice that this is the number of one–one correspondences there are between the elements of the sets

$$\{1, 2, 3\} \quad \text{and} \quad \{✶, ✶, ✶\}$$

But this question is like our first! There are 3 ✶'s which can be labeled 1, and for each way this is done there are 2 ✶'s which can be labeled 2 while the remaining ✶ must accept the label 3. Thus the arrangement

$$✶ \;✶\; 0 \;✶\; 0$$

has been counted

$$3 \cdot 2 = 6$$

times in our count of 60 favorable outcomes. Since the same is true of each of the *distinguishable* arrangements, we actually have only

$$\frac{60}{6} = 10$$

favorable outcomes for the five throws of the die. Thus the probability that numbers less than three will turn up exactly three times in five throws of a die is

$$\frac{4}{243} + \frac{4}{243} + \cdots + \frac{4}{243} = 10 \cdot \frac{4}{243} = \frac{40}{243}$$

The different arrangements of the elements of a set are called *permutations* in mathematics. If the analysis used in Example 9 were applied to the ordered set

$$\{a, b, c, d, e\}$$

(where all five objects are different) we would calculate

$$5 \cdot 4 \cdot 3 \cdot 2 \cdot 1 = 120$$

permutations or arrangements. In considering the permutations obtained from one of these arrangements

$$\{d, b, a, e, c\}$$

merely by interchanging $a$, $b$, and $c$ we would be counting the permutations of

$$\{a, b, c\}$$

of which there are

$$3 \cdot 2 \cdot 1 = 6$$

Now if $a$, $b$, and $c$ were each replaced by a single symbol ✶ then these 6 permutations could not be distinguished. Similarly, if $d$ and $e$ were interchanged, this would yield

$$2 \cdot 1 = 2$$

permutations which could not be distinguished if both $d$ and $e$ were replaced by a single symbol 0. Then for each permutation

$$\{d, b, a, e, c\}$$

of the symbols, $a$, $b$, $c$, $d$, and $e$ exactly

$$[3 \cdot 2 \cdot 1] \cdot [2 \cdot 1] = 6 \cdot 2 = 12$$

of them will look exactly alike, that is, like

$$0 \text{ ✶ ✶ } 0 \text{ ✶}$$

when three of them are alike and the other two are also alike but different from the three. Thus, the number of distinguishable permutations of 5 objects of of which 3 are alike and 2 are alike is given by

$$\frac{5 \cdot 4 \cdot 3 \cdot 2 \cdot 1}{(3 \cdot 2 \cdot 1)(2 \cdot 1)} = \frac{5!}{3!\,2!} = 10$$

Here we have introduced *factorial* notation. Notice that a similar formula applies to the case

$$\text{✶ ✶ } c \text{ ✶ } e$$

where 3 of the 5 objects are alike, while the other two are not:

$$\frac{5!}{3!\,1!\,1!} = \frac{5 \cdot 4 \cdot 3 \cdot 2 \cdot 1}{3 \cdot 2 \cdot 1 \cdot 1 \cdot 1} = 5 \cdot 4 = 20$$

as before, if we interpret 1! as 1. In that case we do not bother to write 1!,

*Special Techniques and Applications*

but simply say that

$$\frac{5!}{3!}$$

represents the number of distinguishable permutations of 5 objects, 3 of which are indistinguishable. When this formula is used for the case where all objects in the set are either of one kind or another, as for ✶ and 0 of Example 9, then it takes the form

$$\frac{n!}{k!\,(n-k)!}$$

It is customary to allow this even for the case

✶ ✶ ✶ ✶ ✶

where all objects are the same. Then we must write

$$\frac{5!}{5!\,(5-5)!} = \frac{5!}{5!\,0!}$$

Since this must equal 1, we accept

$$0! = 1$$

These results are summarized in the following:

*Definition 10.5.*   $n!$ is read *n factorial* where $0! = 1$ and $(n+1)! = (n+1) \cdot n!$ for every $n \in W$.

*Definition 10.6.*   A *permutation* of the elements of a set is a one–one correspondence of the elements with themselves.

## Theorem 10.5

If a finite set $S$ of cardinality $n$ has disjoint subsets of indistinguishable elements of cardinalities $n_1, n_2, \cdots, n_k$, then the number of distinguishable permutations of the elements of $S$ is

$$\frac{n!}{n_1!\,n_2!\cdots n_k!}$$

The situation described by Example 9 can appear in a slightly different form described by the word *combinations*. In our analysis of the distinguishable *permutations* of

✶ ✶ 0 ✶ 0

we made our description in terms of relative positions or *order*. We can accomplish the same analysis by labeling the *positions*

*a b c d e*

and identifying each permutation by the subset of $\{a, b, c, d, e\}$ which represents the positions occupied by the $*$, for example. Thus the set

$$\{a, b, d\}$$

identifies the permutation shown above and

$$\frac{5!}{3!\, 2!}$$

represents the number of distinct subsets of $\{a, b, c, d, e\}$ which are equivalent to $\{a, b, d\}$. In the realm of probability this is described as the *number of combinations of five objects taken three at a time.*

**Definition 10.7.** A *combination* of $k$ elements of a set $S$ of distinct objects where $k \in W$, is a subset $C \subseteq S$ with $n(C) = k$.

# Theorem 10.6

If $S$ is a finite set of $n$ distinct objects, the number of combinations of the elements of $S$ taken $k$ at a time is

$$\frac{n!}{k!\,(n-k)!}$$

where $0 \leq k \leq n$. The notation $\binom{n}{k}$ is often used to represent this number.

## Exercise 10.4

1. Use Definition 10.5 to calculate (a) 1! (b) 2! (c) 3! (d) 4! (e) 7!.

2. The entries of the Pascal Triangle (of numbers) can be found as shown at left by using 1's as first and last entries on each line and adding pairs of numbers from the preceding line as shown. Show that the triangle at right is equivalent.

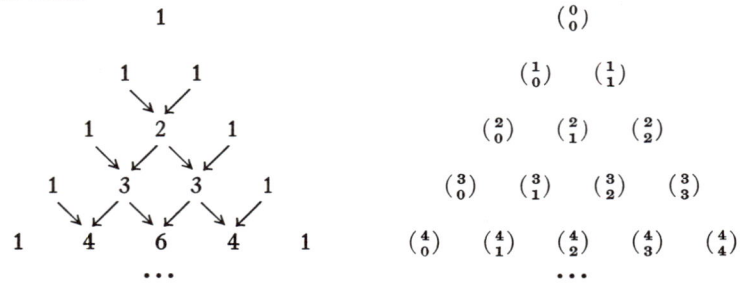

$$
\begin{array}{c}
1 \\
1 \quad 1 \\
1 \quad 2 \quad 1 \\
1 \quad 3 \quad 3 \quad 1 \\
1 \quad 4 \quad 6 \quad 4 \quad 1 \\
\cdots
\end{array}
\qquad
\begin{array}{c}
\binom{0}{0} \\
\binom{1}{0} \quad \binom{1}{1} \\
\binom{2}{0} \quad \binom{2}{1} \quad \binom{2}{2} \\
\binom{3}{0} \quad \binom{3}{1} \quad \binom{3}{2} \quad \binom{3}{3} \\
\binom{4}{0} \quad \binom{4}{1} \quad \binom{4}{2} \quad \binom{4}{3} \quad \binom{4}{4} \\
\cdots
\end{array}
$$

3. A committee of three is to be selected from ten people. How many different committees could be selected?

4. How many different-looking "words" can be made by using all of the letters of M I S S I S S I P P I each time?

*Special Techniques and Applications* 217

5. A chimpanzee takes six cards having the letters E E E P T and arranges them in a row. Assuming that all are upright, what is the probability that he has spelled the word *teepee*?

6. A pair of dice are rolled twice. What is the probability that an 8 occurs first followed by a 7?

7. A coin is flipped ten times. What is the probability that heads shows (a) exactly 3 times? (b) not more than 3 times?

8. If a player wins by rolling a score of either 7 or 11 with one roll of a pair of dice, calculate the probability of such a win by (a) definition (b) compound events.

9. (a) How many different poker hands (5 cards) are there? (b) How many different poker hands of all face cards are there? (c) What is the probability that a poker hand drawn at random will consist of face cards only?

10. Show that if $h + k = n$ then $\binom{n}{h} = \binom{n}{k}$.

11. Verify that

$$\left(\frac{1}{3} + \frac{2}{3}\right)^5 = \binom{5}{5}\left(\frac{1}{3}\right)^5 + \binom{5}{4}\left(\frac{1}{3}\right)^4\left(\frac{2}{3}\right)^1 + \binom{5}{3}\left(\frac{1}{3}\right)^3\left(\frac{2}{3}\right)^2$$

$$+ \binom{5}{2}\left(\frac{1}{3}\right)^2\left(\frac{2}{3}\right)^3 + \binom{5}{1}\left(\frac{1}{3}\right)^1\left(\frac{2}{3}\right)^4 + \binom{5}{0}\left(\frac{2}{3}\right)^5$$

12. The term $\binom{5}{3}(\frac{1}{3})^3(\frac{2}{3})^2$ represents the probability of Example 9 that in five throws of a die, numbers less than 3 will result exactly three times.

$$\binom{5}{3}\left(\frac{1}{3}\right)^3\left(\frac{2}{3}\right)^2 = \frac{5!}{3!(5-3)!}\left(\frac{1}{3}\right)^3\left(\frac{2}{3}\right)^2 = \frac{5\cdot 4\cdot \cancel{3!}}{\cancel{3!}2!}\left(\frac{1}{3}\right)^3\left(\frac{2}{3}\right)^2 = 10\cdot\frac{4}{243}$$

$$= \frac{40}{243}$$

In context of the same situation, what would the terms (a) $\binom{5}{4}(\frac{1}{3})^4(\frac{2}{3})^1$ and (b) $\binom{5}{0}(\frac{2}{3})^5$ represent?

# Selected Answers

*Exercise 1, p. 4*

**1.** Yes. By definition, language is a means of exchanging ideas, and ideas, by definition, are products of the intellect.

**3.** In the exclusive sense, because no one can correctly be classified as both a sophomore and a junior in the same context.

**5.** There is no logical answer to the question.

**9.** In the absence of further information, this sentence is ambiguous. We would probably concede that "Roses are red" means that all roses are red, and therefore, is a false statement.

*Exercise 2, p. 7*

**1.** Let $p$ represent "You prepare your lessons," and $q$ "You fail the course." The only way the statement could be false would be if you passed the course without preparing your lessons.

| $p$ | $q$ | $p \vee q$ |
|---|---|---|
| T | T | T |
| T | F | T |
| F | T | T |
| F | F | F |

**3.** (a) $p \wedge q$  (b) $q \vee \sim p$  (c) $\sim p$  (d) $p \wedge \sim q$

**5.** They have the same truth values.

| $p$ | $q$ | $p \vee q$ | $q \vee p$ |
|---|---|---|---|
| T | T | T | T |
| T | F | T | T |
| F | T | T | T |
| F | F | F | F |

**7.** $\sim(p \vee q)$ is equivalent to $\sim p \wedge \sim q$.

| $p$ | $q$ | $\sim p$ | $\sim q$ | $p \vee q$ | $\sim(p \vee q)$ | $\sim p \vee \sim q$ | $\sim p \wedge \sim q$ |
|---|---|---|---|---|---|---|---|
| T | T | F | F | T | F | F | F |
| T | F | F | T | T | F | T | F |
| F | T | T | F | T | F | T | F |
| F | F | T | T | F | T | T | T |

**9.** When either $p$ or $q$ is false; i.e., in every case except when both are true.

*Exercise 3, p. 12*

**1.** (a) If it is two o'clock, then black is white.
(b) If black is not white, it is two o'clock.
(c) If it is not two o'clock, then black is not white.

**3.** (a) *Hyp.* Black is white; *concl.* It is two o'clock.
(b) Same as (a).
(c) *Hyp.* It is not two o'clock; *concl.* Black is white.

**5.** (a) Black is white if and only if it is two o'clock.
(b) It is two o'clock if and only if black is white.
(c) We hope not!

**8.** Assuming that no error has been made, either you did not pass all your tests or you made less than 90% on your final.

**9.** Either $p$ or $q$ is false.

*Exercise 1.1, p. 17*

**3.** It is unlikely that you will have any equality other than $B = D$.

**5.** $\{*\# \mid *, \#$ are English letters and $*\#$ spells an English word$\}$ This is

*Selected Answers*

probably a great deal easier than poring over an unabridged dictionary. Besides, you may have had to decide which authority was correct!

7. A particular child's set of books, the boys in the room, all the pencils in the teacher's desk, etc.

9. (a) $A = A$ agrees with the concept that each time a symbol is used in a given context, it must represent the same thing. Thus, the two $A$'s must represent the same set, and by Definition 1.1, $A = A$.
(b) By Definition 1.1, both $A = B$ and $B = A$ state that $A$ and $B$ represent the same set. They are logically equivalent statements, each implying the other.
(c) The statements $A = B$ and $B = C$, together, inform us that $A$, $B$, and $C$ represent the same set, so that $A = C$ by Definition 1.1.

10. Call the lists sets $A$ and $B$. See that every name on list $A$ is also on list $B$, so that $A \subseteq B$. Then, to see that $B$ contains no other name, check to see that every name on list $B$ is also on list $A$, or $B \subseteq A$. Thus $A = B$ iff $A \subseteq B$ and $B \subseteq A$.

*Exercise 1.2, p. 21*

1. $a \in A$; $a \in B$; $a \notin C$

3. (a) $\{a, b, d, e, i, o, u\}$ (b) $\{e\}$

5. (a) $\{i, o, u\}$ (b) $\{b, d, r\}$

7. (a) This statement is part of Definition 1.5.
(b) No. If $B = A$, then $A \subseteq B$ but $A \not\subset B$ since $A \subset B$ implies $A \neq B$ by Definition 1.5.

9. (a) If $A = \{a, b, c\}$ and $B = \{a, b\}$, then $A \cup B = \{a, b, c\} = A$, so that $A \not\subset (A \cup B)$. Any case where $B \subseteq A$ is an example.
(b) Same as (a): $A \cap B = \{a, b\} = B$, so $(A \cap B) \not\subset B$.

*Exercise 1.3, p. 24*

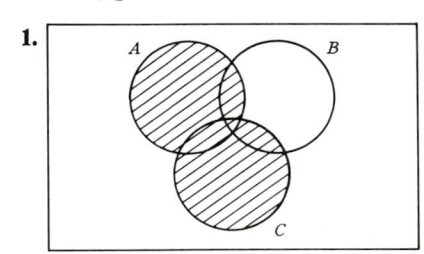

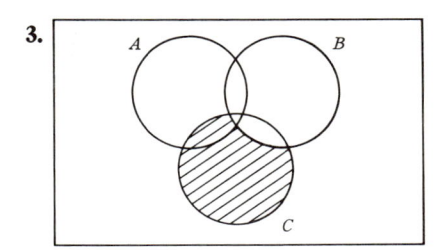

1. [Venn diagram with sets $A$, $B$, $C$; shaded region]   3. [Venn diagram with sets $A$, $B$, $C$; shaded region]

5. $C - A$ or $C \cap A'$

**7.**

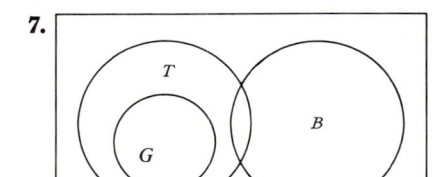

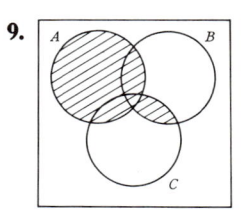

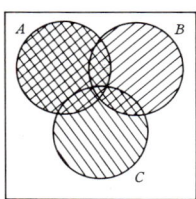

**9.**

**11.** $A - B = A \rightarrow A$ contains no element of $B$, so $A \cap B = \emptyset$.

*Exercise 2.1, p. 29*

1. $C \sim D$, $A \sim C$, $A \sim D$, $B \sim E$

3. $\{(p, \#), (t, \$), (s, *)\}$; $\{(p, \#), (t, *), (s, \$)\}$;
   $\{(p, \$), (t, \#), (s, *)\}$; $\{(p, \$), (t, *), (S, \#)\}$;
   $\{(p, *), (t, \$), (s, \#)\}$; $\{(p, *), (t, \#), (s, \$)\}$.

5. (a) Three; four; three; three; four.
   (b) 3; 4; 3; 3; 4.

7. All are essentially either numerical or alphabetical.

9. $(0, 1), (1, 2), (2, 3), \cdots, (m, n), \cdots$ where $m \in W$ and $n$ follows $m$ in the ordered list $W = \{0, 1, 2, 3, 4, 5, \cdots\}$.

*Exercise 2.2, p. 32*

**1.**

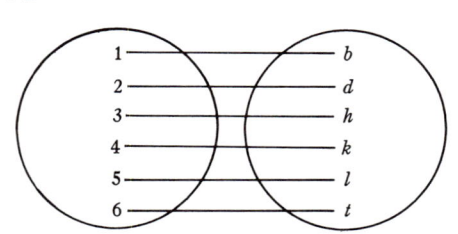

3. (a) $N_4 = \{1, 2, 3, 4\}$; (b) $N_1 = \{1\}$; (c) $N_1 = \{1\}$.

5. (a) $n(A) > n(B)$; (b) $n(A) > n(C)$; (c) $n(D) > n(A)$;
   (d) $n(A) = n(F)$.

7. Let $A = \{a, b, c, d, e, f\}$ and $B = \{x, y, z, w\}$ so that $n(A) = 6$ and $n(B) = 4$. There is a set $C = \{a, b, c, d\}$ such that $C \subset A$, $C \sim B$, and $B \sim A$. Then by Definition 2.5, $n(A) > n(B)$ or $6 > 4$.

Selected Answers

**9.** $n(A) = 3, n(B) = 2, A \cup B = \{a, b, c, d\}$ and $n(A \cup B) = 4$. Since $3 + 2 = 5$ and $5 > 4$, $n(A) + n(B) > n(A \cup B)$.

**11.** $\{4, 6, 8, 10, \cdots\}$, or $\{2, 6, 10, 14, \cdots\}$, or many others.

## Exercise 2.3, p. 36

**1.** $B \cup V = \{b, c, d, a, e, i, o, u\}$; $n(B \cup V) = 8$; $n(B) = 3$; and $n(V) = 5$. Then, since $B \cap V = \emptyset$, $3 + 5 = n(B) + n(V) = n(B \cup V) = 8$.

**3.** If $k \in W$, then there is some set $K$ for which $n(K) = k$ by definition. Also, $n(\emptyset) = 0$ by definition. Then, since $K \cap \emptyset = \emptyset$ and $K \cup \emptyset = K$, $k + 0 = n(K) + n(\emptyset) = n(K \cup \emptyset) = n(K) = k$.

**5.** $2 + 3 = n(A) + n(B) = n(A \cup B) = 5$

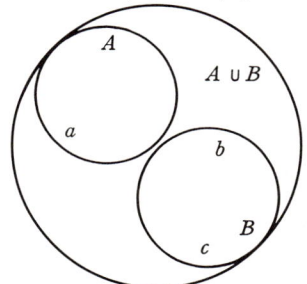

**7.** Let $B = \{b, d, h\}$. Then $n(B) = 3$, $n(A) = 6$, $B \subseteq A$, $A - B = \{a, c, e\}$, and $n(A - B) = 3$, so that $6 - 3 = n(A) - n(B) = n(A - B) = 3$.

**9.** No; only if $B \subseteq A$. Let $A = \{a, b, c, d\}$ and $B = \{d, e, f\}$. Then $A - B = \{a, b, c\}$, $n(A - B) = 3$, $n(B) = 3$, $n(A) = 4$, and $3 + 3 \ne 4$.

## Exercise 3.1, p. 39

**1.** (a) 1,002,135 or one million, two thousand, one hundred thirty-five.
(b) 30,167 or thirty thousand, one hundred sixty-seven.

## Exercise 3.2, p. 40

**1.** (a) CCCXLII
(b) CDXLIX
(c) MDLI

## Exercise 3.3, p. 43

**1.** Baseball or football statistics for a team, telephone number system (area code, exchange, station number).

**3.** Twenty; seven; three hundred; eighty; five; forty-thousand; zero; six hundred; ten; nine.

## Exercise 3.4, p. 45

**1.** (a) Three one four five, base six.
(b) Two zero zero three, base four.

**3.** (a) :) which is in base five and corresponds to $23_{five}$ or 2 fives $+$ 3.
(b) )# represents $34_{five}$ or 3 fives $+$ 4 or nineteen.

**5.** (a) $37_{eight} = 3 \times 8 + 7 = 24 + 7 = 31$.
(b) $50_{seven} = 5 \times 7 + 0 = 35 + 0 = 35$.

## Exercise 3.5, p. 48

**1.** (a) $110_{four}$;  (b) $202_{three}$;  (c) $10100_{two}$

**3.** (a) $312_{four} = 3$ sixteens $+ 1$ four $+ 2$ ones
(b) $2012_{three} = 2$ twenty-sevens $+ 0$ nines $+ 1$ three $+ 2$ ones
(c) $1011_{two} = 1$ sixteen $+ 0$ eights $+ 1$ four $+ 1$ two $+ 0$ ones

**5.** (a) $312_{four} = 3 \times 4^2 + 1 \times 4^1 + 2 \times 4^0$
(b) $2012_{three} = 2 \times 3^3 + 0 \times 3^2 + 1 \times 3^1 + 2 \times 3^0$
(c) $10110_{two} = 1 \times 2^4 + 0 \times 2^3 + 1 \times 2^2 + 1 \times 2^1 + 0 \times 2^0$

**7.** (a) $312_{four} = 3 \times (10_{four})^2 + 1 \times (10_{four})^1 + 2 \times (10_{four})^0$
(b) $2012_{three} = 2 \times (10_{three})^3 + 0 \times (10_{three})^2 + 1 \times (10_{three})^1 + 2 \times (10_{three})^0$
(c) $10110_{two} = 1 \times (10_{two})^4 + 0 \times (10_{two})^3 + 1 \times (10_{two})^2 + 1 \times (10_{two})^1 + 0 \times (10_{two})^0$

## Exercise 3.6, p. 51

**1.** (a) $te_{twelve} = 10 \times 12 + 11 = 120 + 11 = 131$
(b) $21t_{twelve} = 2 \times 144 + 1 \times 12 + 10 = 288 + 12 + 10 = 310$
(c) $4t6_{eleven} = 4 \times 121 + 10 \times 11 + 6 = 484 + 110 + 6 = 600$

**3.** (a) $182_{nine}$;  (b) $11102_{four}$;  (c) $62_{twelve}$

**5.** 1, 2, 3, 10, 11, 12, 13, 20, 21, 22, 23, 30, 31, 32, 33, 100, 101, 102, 103, 110 base four.

## Exercise 4.1, p. 55

**1.** (a) 11, (b) 48, (c) 1001

**3.** If $n \in W$ were the last element, then there would be no $n' \in W$. Since this contradicts Axiom 2, there is no last element in $W$.

**5.** $m + 2 = m + 1' = (m + 1)' = (m')'$

**7.** $2 + 2 = 2 + 1' = (2 + 1)' = (2')' = 3' = 3 + 1$

*Exercise 4.2, p. 58*

1. $E$ and $T$ are closed under addition. In (c), $1 + 3 \notin D$, for example.

3. It is true. Union and intersection are commutative operations on sets.

5. $(A \cap B) \cap C = A \cap (B \cap C)$ is true for all sets $A$, $B$, $C$.

7. (a) $+$ is commutative; (b) $\{0, 2, 4, \cdots\}$ is closed under addition; (c) $+$ is associative; (d) $+$ is commutative; (e) $+$ is commutative *and* the definition of $=$.

9. $a = a_1$ is the trivial case which follows from the hypothesis. If $a = a_1$ and $a_1 = a_2$ then $a = a_2$ by the transitive property of $=$. Suppose, for some $k \in N$, that $a = a_1, a_1 = a_2, \cdots, a_{k-1} = a_k$ implies $a = a_k$; then if $a = a_1, a_1 = a_2, \cdots, a_{k-1} = a_k$, and $a_k = a_{k+1}$, it follows that $a = a_k$ which, with $a_k = a_{k+1}$, implies $a = a_{k+1}$.

*Exercise 4.3, p. 64*

1. All three sets.

3. For every $A$, $B$ and $C$, $A \cap (B \cup C) = (A \cap B) \cup (A \cap C)$ is true. Argument: $x \in A \cap (B \cup C)$ implies $x \in A$ and either $x \in B$ or $x \in C$; that is, either $x \in A$ and $x \in B$ or $x \in A$ and $x \in C$, which means $x \in A \cap B$ or $x \in A \cap C$, so that $x \in (A \cap B) \cup (A \cap C)$. This shows $A \cap (B \cup C) \subseteq (A \cap B) \cup (A \cap C)$. A similar analysis which you should make shows that every element of $(A \cap B) \cup (A \cap C)$ is also an element of $A \cap (B \cup C)$, or $(A \cap B) \cup (A \cap C) \subseteq A \cap (B \cup C)$ which, with $A \cap (B \cup C) \subseteq (A \cap B) \cup (A \cap C)$, implies that the two symbols represent the same set.

5. (a) $+$ is associative; (b) $\times$ is distributive over $+$ (perhaps also $\times$ is commutative and definition of $=$); (c) $\times$ is commutative; (d) $\times$ is distributive over $+$; (e) $\times$ is associative.

7. $3 \times 5 = 3 \times (2 + 3) = (3 \times 2) + (3 \times 3)$.

9. $1 \times 0 = 0$ by Definition 4.2. If $1 \times k = k$, then $1 \times k' = 1 \times k + 1 = k + 1 = k'$ by Definition 4.1. By induction, $1 \times n = n$ for $n \in W$.

*Exercise 5.1, p. 71*

1. (a) $T = \{0, 10, 20\}$; (b) $T = \{2, 4, 6, 8, \cdots\}$; (c) $T = \{\text{dime, quart}\}$

3. (a), (b), (c) and (e)

5. (a) $A - B = A$ and $B - A = B$, so $A - B \neq B - A$.
   (b) $(C - A) - B = B - B = \emptyset$, while $C - (A - B) = C - A = B$.

7. Since $A \cup B = B \cup A$ for all sets $A$ and $B$, by Definition 2.6, $n(A) + n(B) = n(A \cup B) = n(B \cup A) = n(B) + n(A)$ when $A \cap B = \emptyset$.

9. (a) $W$ is closed under $+$; (b) $+$ is commutative; (c) $+$ is associative.

*Selected Answers*

*Exercise 5.2, p. 78*

**1.** (a) 3,075 or three thousand seventy-five
(b) 6,254 or six thousand two hundred fifty-four.

**3.** (a) base five; (b) $2013_{\text{five}}$; (c) two hundred fifty-eight.

**5.** (2 thirty-sixes + 1 six + 0 ones) + (1 thirty-six + 3 sixes + 5 ones) = (2 + 1) thirty-sixes + (1 + 3) sixes + (0 + 5) ones = 3 thirty-sixes + 4 sixes + 5 ones = $345_{\text{six}}$.

**7.**
(a)  32
    $13_{\text{six}}$
    ―――
     5
    $40_{\text{six}}$
    ―――
    $45_{\text{six}}$

(b)  304
    $123_{\text{six}}$
    ―――
     11
     20
    $400_{\text{six}}$
    ―――
    $431_{\text{six}}$

(c)  303
    $253_{\text{six}}$
    ―――
     10
     50
    $500_{\text{six}}$
    ―――
    100
    $500_{\text{six}}$
    ―――
    $1000_{\text{six}}$

**9.**

| + | 0 | 1 | 2 | 3 | 4 |
|---|---|---|---|---|---|
| 0 | 0 | 1 | 2 | 3 | 4 |
| 1 | 1 | 2 | 3 | 4 | 10 |
| 2 | 2 | 3 | 4 | 10 | 11 |
| 3 | 3 | 4 | 10 | 11 | 12 |
| 4 | 4 | 10 | 11 | 12 | 13 |

*Exercise 5.3, p. 81*

**1.** (a) $\{a, e\}$; (b) $\{b, d\}$; (c) $\{c, f\}$

**3.** Point $d$ is farther. Point selected should be on another circle with $b$ as center with does not intersect the given circle at any other point.

**5.** Use the same radius to draw circles with centers $b$ and $c$. If they intersect, this locates one or two points; if not, use a larger radius. Repeat with different radii.

*Exercise 5.4, p. 83*

**1.** $a(b)c$, $a(b)d$, $a(c)d$, $b(c)d$

**3.** Draw circle with center $a$ and radius $|\,cd\,|$, and another with center $b$ and radius $|\,ef\,|$. The two circles intersect in two points which are the elements of the set.

5. $\overset{\cdot}{a}\ \overset{\circ\longrightarrow}{b}$ The small circle in place of a dot at $b$ indicates that $b$ is not a point in the set (neither is $a$).

## Exercise 5.5, p. 87

3. (a) $\overline{ab}$; (b) $\overleftrightarrow{ab}$; (c) triangle $acd$; (d) angle $cbd$; (e) $\overrightarrow{ab}$

## Exercise 6.1, p. 94

1. (a) $A \times B = \{(a, b), (a, c), (b, b), (b, c)\}$; $B \times A = \{(b, a), (b, b), (c, a), (c, b)\}$.
   (b) $A \times B = \{(0, 0), (0, 1), (0, 2), (1, 0), (1, 1)(1, 2), (2, 0), (2, 1), (2, 2)\} = B \times A$.

3. $1 \cdot 3$ described as *one three* might be adequate, but describing $0 \cdot 3$ as *zero threes* is undesirable.

5. $W$ is closed under both $+$ and $\cdot$; $+$ and $\cdot$ are associative; $+$ and $\cdot$ are commutative; $\times$ is associative over $+$; 0 is an identity for $+$; 1 is an identity for $\times$.

7. (a) "Yes" to both questions. (b) Set union is distributive over set intersection.

9. (a) $\times$ is associative; (b) $\times$ is distributive over $+$; (c) $+$ is commutative; (d) $\times$ is distributive over $+$; (e) $+$ is associative; (f) $\times$ is commutative; (g) $\times$ is distributive over $+$; (h) $+$ is commutative.

## Exercise 6.2, p. 100

1. (a) $32 \times 23 = 32 \times (2 \times 10 + 3) = 32 \times (2 \times 10) + 32 \times 3 = (3 \times 10 + 2) \times (2 \times 10) + (3 \times 10 + 2) \times 3 = (3 \times 10) \times (2 \times 10) + 2 \times (2 \times 10) + (3 \times 10) \times 3 + 2 \times 3 = (3 \times 2) \times (10 \times 10) + (2 \times 2) \times 10 + (3 \times 3) \times 10 + 2 \times 3 = 6 \times 100 + 4 \times 10 + 9 \times 10 + 6 = 6 \times 100 + (4 + 9) \times 10 + 6 = 6 \times 100 + 13 \times 10 + 6 = 6 \times 100 + (1 \times 10 + 3) \times 10 + 6 = 6 \times 100 + (1 \times 10) \times 10 + 3 \times 10 + 6 = 6 \times 100 + 1 \times 100 + 3 \times 10 + 6 = (6 + 1) \times 100 + 3 \times 10 + 6 = 7 \times 100 + 3 \times 10 + 6 = 736$
   (b) $12_{\text{seven}} \times 32_{\text{seven}} = 12_{\text{seven}} \times (3 \times 7 + 2) = 12_{\text{seven}} \times (3 \times 7) + 12_{\text{seven}} \times 2 = (1 \times 7 + 2) \times (3 \times 7) + (1 \times 7 + 2) \times 2 = (1 \times 7)(3 \times 7) + 2 \times (3 \times 7) + (1 \times 7) \times 2 + 2 \times 2 = (1 \times 3) \times (7 \times 7) + (2 \times 3) \times 7 + (1 \times 2) \times 7 + 2 \times 2 = 3 \times 49 + 6 \times 7 + 2 \times 7 + 4 = 3 \times 49 + (6 + 2) \times 7 + 4 = 3 \times 49 + 11_{\text{seven}} \times 7 + 4 = 3 \times 49 + (1 \times 7 + 1) \times 7 + 4 = 3 \times 49 + (1 \times 7) \times 7 + 1 \times 7 + 4 = 3 \times 49 + 1 \times (7 \times 7) + 1 \times 7 + 4 = (3 + 1) \times 49 + 1 \times 7 + 4 = 4 \times 49 + 1 \times 7 + 4 = 414_{\text{seven}}$

**3.** (c)

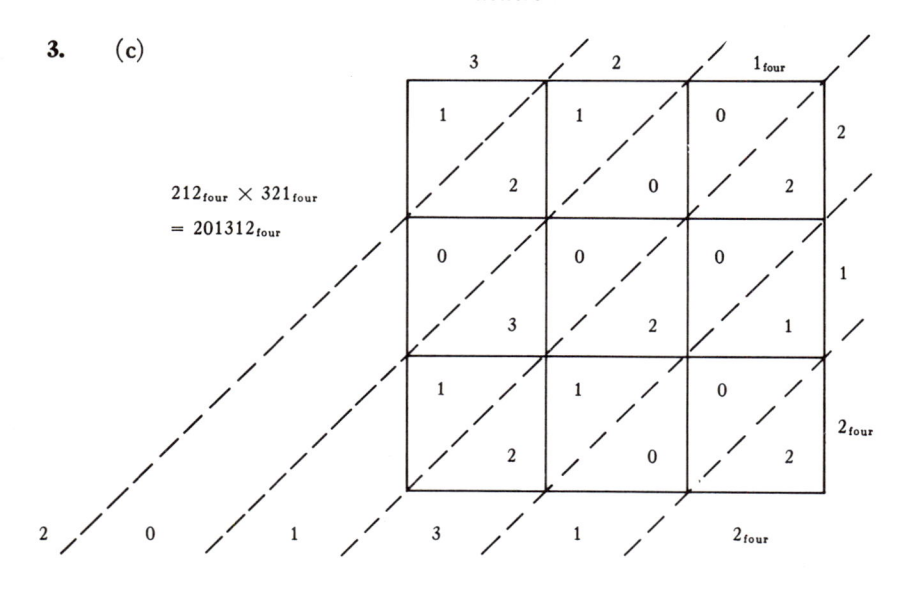

$212_{four} \times 321_{four}$
$= 201312_{four}$

**5.** (a) $2013_{four}$; (b) $21102_{four}$

**7.**

| + | 0 | 1 | 2 | 3 | 4 |
|---|---|---|---|---|---|
| 0 | 0 | 1 | 2 | 3 | 4 |
| 1 | 1 | 2 | 3 | 4 | 10 |
| 2 | 2 | 3 | 4 | 10 | 11 |
| 3 | 3 | 4 | 10 | 11 | 12 |
| 4 | 4 | 10 | 11 | 12 | 13 |

**9.** (a) $11,240_{five}$; (b) $21,344_{five}$

*Exercise 6.3, p. 106*

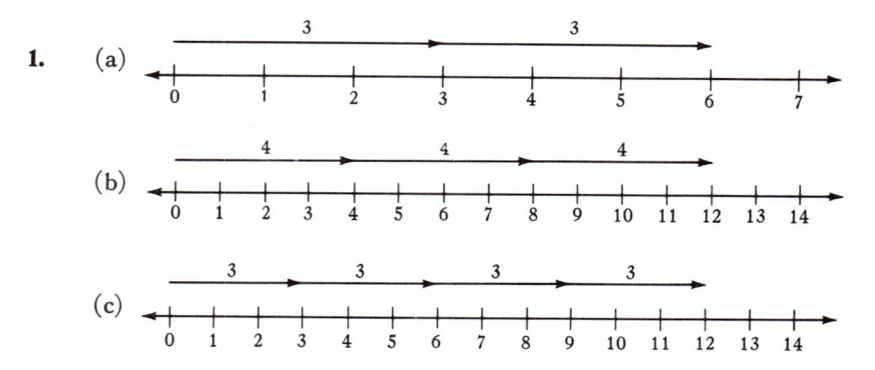

**1.** (a)

(b)

(c)

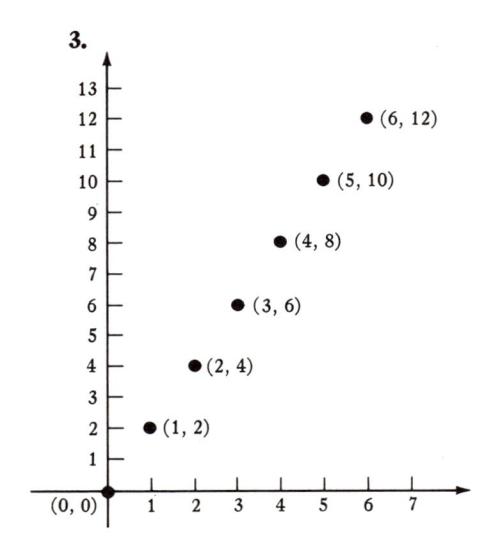

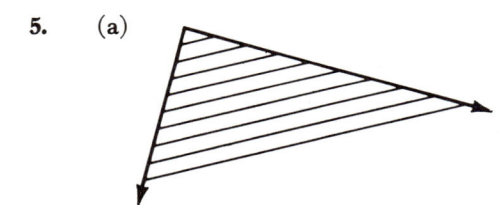

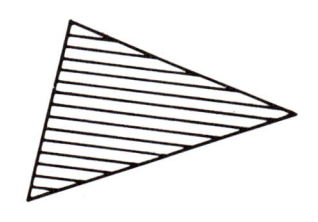

**3.**

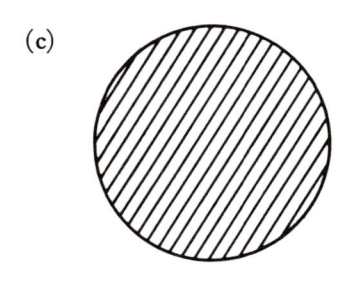

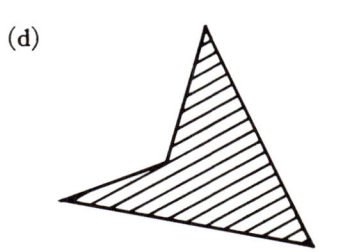

[Graph showing points (0,0), (1,2), (2,4), (3,6), (4,8), (5,10), (6,12)]

**5.** (a) (b) (c) (d)

**7.**   (c) and (d)

**9.**   (a) $108 \times 4 = 432$;   (b) $108 \times 9 = 972$

## Exercise 7.1, p. 114

**1.**   (a) $48 - 15 = (4 \times 10 + 8) - (1 \times 10 + 5) = (4 \times 10 - 1 \times 10) + (8 - 5) = (4 - 1) \times 10 + (8 - 5) = 3 \times 10 + 3 = 33$

## Selected Answers

(b) $713 - 246 = (7 \times 100 + 1 \times 10 + 3) - (2 \times 100 + 4 \times 10 + 6) = (6 \times 100 + 10 \times 10 + 13) - (2 \times 100 + 4 \times 10 + 6) = (6 - 2) \times 100 + (10 - 4) \times 10 + (13 - 6) = 4 \times 100 + 6 \times 10 + 7 = 467$

3.  (c) $\overset{7\ 11}{\cancel{8}\cancel{1}}$  (d) $\overset{6\ \ 10\ \ 13}{\cancel{7}\cancel{1}\cancel{3}}$
    $\underline{-45}$     $\underline{-246}$
    $\ \ \ 36$     $\ \ \ 467$

5.  By Definition 7.1, $m - n = k$ iff $k + n = m$ where $m \geq n$. Thus for $m \geq n$ there is $k \in W$ for which $m - n = k$, so that by Definition 7.1, $k + n = m$ or $(m - n) + n = m$.

7.  (a) $n = 11 - 5$; (b) $n = 11 - 5$; (c) $n = 13 + 6$; (d) $n = 13 - 9$; (e) $n = (14 + 4) - 7$

9.  (a) 6; (b) 12; (c) 6; (d) undefined unless a convention is adopted. We will use the convention that when numbers are listed without punctuation on a row, with one operation or an operation and its inverse to be performed, the numbers will be combined in a cumulative manner from left to right. (See Problem 10)

11. $a - b - c = (a - b) - c = a - (b + c)$ because $(a - b - c) + (b + c) = \{[(a - b) - c] + c\} + b = (a - b) + b = a$.

## Exercise 7.2, p. 120

1.  (a) (c)
```
* * * * * *        * * *
* * * * * *   or   * * *
* * * * * *        * * *
                   * * *
                   * * *
                   * * *
```

3.  (a) 2, 2; (b) 13, 13; (c) 119, 119

5.  (a) $n = 56 \div 8$; (b) $n = 9 \times 4$; (c) $n = 0 \div 6$; (d) $n$ is undefined; (e) $n = (20 \div 5) - 3$

7.  (a) $3400 \div 17 = (34 \times 100) \div 17 = (34 \div 17) \times 100 = 2 \times 100 = 200$
    (b) $(63{,}000 + 2700 + 450) \div 9 = (63{,}000 \div 9) + (2700 \div 9) + (450 \div 9) = [(63 \times 1000) \div 9] + [(27 \times 100) \div 9] + [(45 \times 10) \div 9] = [(63 \div 9) \times 1000] + [(27 \div 9) \times 100] + [(45 \div 9) \times 10] = 7 \times 1000 + 3 \times 100 + 5 \times 10 + 0 \times 1 = 7350$

*Selected Answers*

(c) $360 \div 8 = (320 + 40) \div 8 = [(32 \times 10) \div 8] + (40 \div 8) = (32 \div 8) \times 10 + (40 \div 8) = 4 \times 10 + 5 = 45$

9. (a) $66{,}150 = 90(735) + 0$; (b) $34{,}162 = 273(125) + 37$

11. (1) $n \div n = 1$ because $1 \times n = n$.
(2) If $m \div n = k \in W$; then $k \times n = m$ or $(m \div n) \times n = m$.
(3) $h \cdot n = k \cdot n \rightarrow h \cdot n - k \cdot n = 0 \rightarrow (h - k) \cdot n = 0 \rightarrow h - k = 0$, or $h = k$.
(4) If $(m \cdot n) \div n = k \in W$, then $k \times n = (m \cdot n)$ which implies $k = m$, or $(m \cdot n) \div n = m$.

## Exercise 7.3, p. 126

1. $C = \{[(a, x), (b, y), (c, z)], [(d, x), (e, y), (f, z)], [(g, x), (h, y), (i, z)], [(j, x), (k, y), (m, z)]\}$, or $\{(a, b, c), (d, e, f), (g, h, i), (j, k, m)\} = C$.

3. (a) $4_{six}$; (b) $5_{six}, 5_{six}$; (c) $14_{six}$; (d) $3054_{six}$

5. (a) $345_{six} - 132_{six} = (3 \times 36 + 4 \times 6 + 5) - [(1 \times 36) + 3 \times 6 + 2] = (3 - 1) \times 36 + (4 - 3) \times 6 + (5 - 2) = 2 \times 36 + 1 \times 6 + 3 = 213_{six}$.
$345_{six} - 132_{six} = (300_{six} + 40_{six} + 5_{six}) - (100_{six} + 30_{six} + 2_{six}) = (300 - 100)_{six} + (40 - 30)_{six} + (5 - 2)_{six} = 200_{six} + 10_{six} + 3_{six} = 213_{six}$.
(b) $432_{six} - 145_{six} = (4 \times 36 + 3 \times 6 + 2) - (1 \times 36 + 4 \times 6 + 5) = (3 \times 36 + 10_{six} \times 6 + 3 \times 6 + 2) - (1 \times 36 + 4 \times 6 + 5) = (3 \times 36 + 13_{six} \times 6 + 2) - (1 \times 36 + 4 \times 6 + 5) = (3 \times 36 + 12_{six} \times 6 + 10_{six} + 2) - (1 \times 36 + 4 \times 6 + 5) = (3 \times 36 + 12_{six} \times 6 + 12_{six} \times 1) - (1 \times 36 + 4 \times 6 + 5 \times 1) = (3 - 1) \times 36 + (12_{six} - 4) \times 6 + (12_{six} - 5) \times 1 = 2 \times 26 + 4 \times 6 + 3 \times 1 = 243_{six}$.
$432_{six} - 145_{six} = (400_{six} + 30_{six} + 2) - (100_{six} + 40_{six} + 5) = (300_{six} + 100_{six} + 20_{six} + 10_{six} + 2) - (100_{six} + 40_{six} + 5) = (300 - 100)_{six} + (120_{six} - 40_{six}) + (12_{six} - 5) = 200_{six} + 40_{six} + 3_{six} = 243_{six}$.

7. (a)  15
$\phantom{0}$ $-3$  1
$\phantom{00}$ ―――
$\phantom{00}$ 12
$\phantom{0}$ $-3$  2
$\phantom{00}$ ―――
$\phantom{000}$ 9
$\phantom{0}$ $-3$  3  $\quad 15 \div 3 = 5$
$\phantom{00}$ ―――
$\phantom{000}$ 6
$\phantom{0}$ $-3$  4
$\phantom{00}$ ―――
$\phantom{000}$ 3
$\phantom{0}$ $-3$  5
$\phantom{00}$ ―――
$\phantom{000}$ 0

(b)  18
$\phantom{0}$ $-6$  1
$\phantom{00}$ ―――
$\phantom{00}$ 12
$\phantom{0}$ $-6$  2 $\quad 18 \div 6 = 3$
$\phantom{00}$ ―――
$\phantom{000}$ 6
$\phantom{0}$ $-6$  3
$\phantom{00}$ ―――
$\phantom{000}$ 0

9.  (a) 0, 0; (b) impossible since $n \times 0 = 0$ for every number $n$; (c) $0 \div 0$ does not define a unique number even if it were not excluded by Definition 7.1, since $n \times 0 = 0$ for *every* number $n$.

11. (a) $34_{six} \overline{)1422_{six}}$ quotient $25_{six}$; 112, 302, 302     (b) $512_{six}$

    (c) $15304_{six} = 312_{six}(33_{six}) + 124_{six}$

*Exercise 8.1, p. 134*

1.  (a) $x = 12$; (b) $x = 0$; (c) $x = 0$; (d) $x = 0$; (e) $y = 7$

3.  (a) $\{0, 1, 2, 3\}$; (b) $\{1\}$; (c) $\{\ \} = \emptyset$

5.  (a), (b), (d), (e)

7.  (a) [number line from $-10$ to $0$ showing $-5$ and $-4$]

    (d) [number line from $-7$ to $0$ showing $5$]

9.  (a) $x = 7$; (b) $y = 0$; (c) $x = {}^-8$; (d) $n = {}^-5$; (e) $n = {}^-11$; (f) $n = 4$

*Exercise 8.2, p. 139*

1.  (a) ${}^-8 + {}^-9 = {}^-(8 + 9) = {}^-17$; (b) ${}^-11 + 4 = {}^-(11 - 4) = {}^-7$; (c) $5 + {}^-3 = 5 - 3 = 2$; (d) ${}^-1 + {}^-4 = {}^-(1 + 4) = {}^-5$; (e) $17 + {}^-6 = 17 - 6 = 11$; (f) $10 + {}^-19 = {}^-(19 - 10) = {}^-9$

3.  (a) $2 \times {}^-6 = {}^-6 + {}^-6 = {}^-(6 + 6) = {}^-(2 \times 6)$
    (b) $4 \times {}^-5 = {}^-5 + {}^-5 + {}^-5 + {}^-5 = {}^-10 + {}^-10 = {}^-20 = {}^-(4 \times 5)$
    (c) ${}^-3 \times 4 = 4 \times {}^-3 = {}^-3 + {}^-3 + {}^-3 + {}^-3 = {}^-(3 + 3 + 3 + 3) = {}^-(4 \times 3) = {}^-(3 \times 4)$

5.  $b = b * e = b * (c * d) = (b * c) * d = e * d = d$

7.  (a) $11 - {}^-3 = 11 + {}^-({}^-3) = 11 + 3 = 14$; (b) ${}^-3 - 11 = {}^-3 + {}^-11 = {}^-14$; (c) $0 - 6 = 0 + {}^-6 = {}^-6$; (d) ${}^-4 - 0 = {}^-4 + {}^-0 = {}^-4 + 0 = {}^-4$;

*Selected Answers*

(e) $^-6 - {}^-5 = {}^-6 + {}^-({}^-5) = {}^-6 + 5 = {}^-1$;  (f) $^-9 - {}^-9 = {}^-9 + {}^-({}^-9) = {}^-9 + 9 = 0$

9. (a) $12 \div {}^-4 = {}^-(12 \div 4) = {}^-3$;  (b) $^-18 \div {}^-3 = 18 \div 3 = 6$;  (c) $^-20 \div 4 = {}^-(20 \div 4) = {}^-5$;  (d) $0 \div {}^-6 = {}^-(0 \div 6) = {}^-0 = 0$

11. Need only prove $1 \times n = n$ for $n \in \{{}^-1, {}^-2, {}^-3, \cdots\}$. Then there is $m \in N$ for which $n = {}^-m$. Then by Theorem 8.3, $1 \times n = 1 \times {}^-m = {}^-(1 \times m) = {}^-m = n$.

13. (a) $^-4 \times {}^-3 = 4 \times 3 = 12$;  (b) $6 \times {}^-3 = {}^-(6 \times 3) = {}^-18$;  (c) $^-5 \times 4 = {}^-(5 \times 4) = {}^-20$;  (d) $0 \times {}^-6 = 0$

## Exercise 8.3, p. 145

1. (a) $\overline{{}^-1} = 6$;  (b) $\overline{{}^-2} = 5$;  (c) $\overline{{}^-3} = 4$;  (d) $\overline{{}^-4} = 3$;  (e) $\overline{{}^-5} = 2$;  (f) $\overline{{}^-6} = 1$

3. (a) $3 \times \overline{4} = 3 \times 2 = 6$;  (b) $(2 \times \overline{6}) \times (5 \times \overline{4}) = (2 \times 6) \times (5 \times 2) = 5 \times 3 = 1$;  (c) $3 \times 5 + 4 \times 2 = 3 \times 3 + 4 \times 4 = 2 + 2 = 4$

5. (a) $5 \times \overline{4} = 5 \times 2 = 3$, and $4 \times \overline{6} = 4 \times 6 = 3$ also.
   (b) $5 \times 6 = 2$ and $4 \times 4 = 2$ also.

7. $(c \times \overline{d}) \times (d \times \overline{c}) = (c \cdot d) \times \overline{d \cdot c} = 1 = 1 \times \overline{1}$ since $c \cdot d \cdot 1 = d \cdot c \cdot 1$

9. (a) $0 \times \overline{b} = 0 \times \overline{c}$ since $0 \cdot c = 0 = b \cdot 0$
   (b) If $a \times \overline{b} \in F$ then $0 \oplus (a \times \overline{b}) = (0 \times \overline{1}) \oplus (a \times \overline{b}) = (0 \cdot b + 1 \cdot a) \times \overline{1 \cdot b} = (0 + a) \times \overline{1 \cdot b} = a \times \overline{b}$

11. (a) $m \times \overline{1} \oplus n \times \overline{1} = (m \cdot 1 + 1 \cdot n) \times \overline{1 \cdot 1} = (m + n) \times \overline{1} = m + n$
    (b) $m \times \overline{1} \otimes n \times \overline{1} = (m \cdot n) \times \overline{1 \cdot 1} = (m \cdot n) \times \overline{1} = m \cdot n$

## Exercise 8.4, p. 152

1. (a) $\dfrac{{}^-5}{8}$;  (b) $\dfrac{{}^-3}{6}$;  (c) $\dfrac{6}{2}$;  (d) $\dfrac{{}^-4}{6}$;  (e) $\dfrac{9}{6}$

3. (a) $\dfrac{{}^-4}{3}$;  (b) $\dfrac{1}{1}$;  (c) $\dfrac{{}^-1}{3}$;  (d) $\dfrac{7}{4}$;  (e) $\dfrac{0}{1}$

5. (a) 5;  (b) $^-4$;  (c) 0;  (d) $^-21$;  (e) undefined

7. (a) 14;  (b) 1;  (c) 9

9. (a) $\dfrac{{}^-5}{13}$;  (b) $\dfrac{{}^-14}{28} = \dfrac{{}^-1}{2}$;  (c) $\dfrac{10}{15} = \dfrac{2}{3}$;  (d) $\dfrac{10}{15} = \dfrac{2}{3}$

## Exercise 8.5, p. 160

1. (a) $4 + \dfrac{13}{15} = \dfrac{60}{15} + \dfrac{13}{15} = \dfrac{73}{15}$; (b) $\left(2 + \dfrac{1}{8}\right) + \left(3 + \dfrac{3}{8}\right) = (2 + 3) + \left(\dfrac{1}{8} + \dfrac{3}{8}\right) = 5 + \dfrac{4}{8} = 5 + \dfrac{1}{2} = \dfrac{10}{2} + \dfrac{1}{2} = \dfrac{11}{2}$; (c) $\dfrac{22}{14} = \dfrac{11}{7}$; (d) $\dfrac{95}{12}$

3. (a) $2 \cdot 2 \cdot 2 \cdot 3 \cdot 11$; (b) $3 \cdot 3 \cdot 5 \cdot 13$; (c) $3 \cdot 3 \cdot 3 \cdot 3 \cdot 7 \cdot 7$; (d) $7 \cdot 17 \cdot 19$

5. (a) $\dfrac{5}{12} + \dfrac{2}{3} = \dfrac{5}{12} + \dfrac{8}{12} = \dfrac{13}{12}$; (b) $\dfrac{9(3)}{7 \cdot 13(3)} + \dfrac{^-2(7)}{3 \cdot 13(7)} = \dfrac{27 - 14}{3 \cdot 7 \cdot 13} = \dfrac{1}{21}$;

   (c) $\dfrac{23}{246}$

7. (a) $\dfrac{^-11}{42}$; (b) $\dfrac{^-1}{21}$; (c) $\dfrac{^-182}{3 \cdot 3 \cdot 7 \cdot 13} = \dfrac{^-2}{9}$

9. (a) $^-3 - {^-7} = {^-3} + 7 = 4 > 0$, so $^-3 > {^-7}$; (b) 2; (c) 0; (d) $\dfrac{^-3}{4}$;

   (e) $\dfrac{25}{57}$; (f) $\dfrac{0}{^-7} = 0$

11. (a) Since $p - 0 = p$ and $p$ is positive, $p > 0$ by Definition 8.10.
    (b) If $n$ is negative, $0 - n = 0 + {^-n} = {^-n}$ which is positive by Problem 10, so that $0 > n$ by Definition 8.10.

## Exercise 9.1, p. 168

1. (a) $2 \times 100 + 3 \times 10 + 4 \times 1 + 5 \times \dfrac{1}{10} + 6 \times \dfrac{1}{100}$

   (b) $5 \times 10 + 0 \times 1 + 0 \times \dfrac{1}{10} + 0 \times \dfrac{1}{100} + 4 \times \dfrac{1}{1000}$

   (c) $0 \times 1 + 0 \times \dfrac{1}{10} + 0 \times \dfrac{1}{100} + 3 \times \dfrac{1}{1000} + 0 \times \dfrac{1}{10,000} + 7 \times \dfrac{1}{100,000}$

*Selected Answers*

3. (a) $3 \times 6 + 2 \times 1 + 4 \times \frac{1}{6} + 2 \times \frac{1}{36} + 1 \times \frac{1}{216}$

   (b) $1 \times 8 + 1 \times 4 + 0 \times 2 + 1 \times 1 + 0 \times \frac{1}{2} + 1 \times \frac{1}{4} + 0 \times \frac{1}{8} + 1 \times \frac{1}{16}$

   (c) $0 \times 1 + 0 \times \frac{1}{4} \times 1 \times \frac{1}{16} \times 2 \times \frac{1}{64} + 3 \times \frac{1}{256}$

5. (a) $\left(2 \times 10 + 0 \times 1 + 1 \times \frac{1}{10} + 3 \times \frac{1}{100}\right) + \left(7 \times 1 + 2 \times \frac{1}{10} + 5 \times \frac{1}{100} + 4 \times \frac{1}{1000}\right) = 2 \times 10 + (0 + 7) \times 1 + (1 + 2) \times \frac{1}{10} + (3 + 5) \times \frac{1}{100} + 4 \times \frac{1}{1000} = 2 \times 10 + 7 \times 1 + 3 \times \frac{1}{10} + 8 \times \frac{1}{100} + 4 \times \frac{1}{1000} = 27.384$

7. (a) $\frac{234}{100} = \frac{117}{50}$; (b) $\frac{73}{100}$; (c) $\frac{1,007}{1,000}$; (d) $\frac{305}{10,000} = \frac{61}{2,000}$

9. (a) $\frac{203}{10} \times \frac{27}{10} = \frac{203 \times 27}{100} = \frac{5,481}{100} = 54.81$; (b) $\frac{23}{1000} \times \frac{407}{100} = \frac{9361}{100,000} = 0.09361$; (c) $0.00063$

11. (a) $107,041 \div 4,070$; (b) $43,000 \div 17$; (c) $60 \div 123$

13. (a)
```
          26.3
   4.07 ) 107.041
          81 4
          ─────
          25 64
          24 42
          ─────
           1 221
           1 221
```
(b) $3.07$

*Exercise 9.2, p. 175*

1. (a) $0.7272\cdots = 0.\overline{72}$; (b) $2.6\overline{6}\cdots$; (c) $0.\overline{24}24\cdots$; (d) $1.\overline{142857}$; (e) $0.\overline{619047}$; (f) $0.\overline{1764705882352941}$

3. (a) $\frac{2}{3}$; (b) $\frac{3}{11}$; (c) $\frac{23}{37}$; (d) $\frac{^-410}{111}$

**5.** (a) $\dfrac{1}{6} = \dfrac{8 \cdot \frac{1}{6}}{8} = \dfrac{1\frac{1}{3}}{8} = \dfrac{1}{8} + \dfrac{\frac{1}{3}}{8} = \dfrac{1}{8} + \dfrac{8 \cdot \frac{1}{3}}{8^2} = \dfrac{1}{8} + \dfrac{2\frac{2}{3}}{8^2} = \dfrac{1}{8} + \dfrac{2}{8^2} + \dfrac{\frac{2}{3}}{8^2} = \dfrac{1}{8} +$

$\dfrac{2}{8^2} + \dfrac{8 \cdot \frac{2}{3}}{8^3} = \dfrac{1}{8} + \dfrac{2}{8^2} + \dfrac{5\frac{1}{3}}{8^3} = \dfrac{1}{8} + \dfrac{2}{8^2} + \dfrac{5}{8^3} + \dfrac{\frac{1}{3}}{8^3} = 0.12525\cdots_{\text{eight}}$;

(b) $0.1\overline{444}\cdots_{\text{nine}}$;  (c) $0.0\overline{111}\cdots_{\text{three}}$;  (d) $0.00\overline{101}\cdots_{\text{two}}$

**7.** (a) $33_{\text{four}} \times 0.\overline{1313}\cdots_{\text{four}} = (100_{\text{four}} - 1) \times 0.\overline{1313}\cdots_{\text{four}} = 13.\overline{1313}\cdots_{\text{four}} -$

$0.\overline{1313}\cdots_{\text{four}} = 13_{\text{four}}$, so that $0.\overline{1313}\cdots_{\text{four}} = \dfrac{13_{\text{four}}}{33_{\text{four}}} = \left(\dfrac{13}{33}\right)_{\text{four}}$

(b) $\left(\dfrac{3}{5}\right)_{\text{six}}$;  (c) $\left(\dfrac{7}{7}\right)_{\text{eight}} = 1$;  (d) $\left(\dfrac{101}{111}\right)_{\text{two}}$

**9.** Since $p$ is in the left half of the interval $[0, 1)$ the first digit is 0. $p$ is also in the left half of the interval $[0, 0.1)$, so the second digit is also zero. In the next bisection, $p$ falls in the interval $[0.001, 0.01)$ so the third is 1, while in the fourth bisection it falls in $[0.0011, 0.01)$. This gives $1/5 = 0.0011\cdots_{\text{two}}$ where the 1 does not necessarily (from our analysis) continue to repeat. Actually,

$0.\overline{0011}_{\text{two}} = \dfrac{11_{\text{two}}}{1111_{\text{two}}} = \dfrac{3}{15} = \dfrac{1}{5}$

**11.** (a) $\dfrac{41}{333} + \dfrac{410}{333} = \dfrac{451}{333} = 1.\overline{354}$  (b) $\dfrac{410}{333} - \dfrac{41}{333} = \dfrac{369}{333} = \dfrac{41}{37} = 1.\overline{108}$

(c) $9 \times \dfrac{123}{999} = \dfrac{123}{111} = \dfrac{41}{37} = 1.\overline{108}$  (d) $\dfrac{8}{11} \div \dfrac{2}{3} = \dfrac{8}{11} \cdot \dfrac{3}{2} = \dfrac{12}{11} = 1.\overline{09}$

## Exercise 9.3, p. 184

**1.** If $n$ is a positive rational number, then there are natural numbers $p$ and $q$ for which $n = p/q$, so that $n^2 = 3$ implies that $p^2/q^2 = 3$ or $p^2 = 3q^2$. Each of $p^2$ and $q^2$ is a product of an even number of prime factors, so that $3 \cdot q^2$ is a product of an odd number of prime factors. But $p^2 = 3q^2$ means that $p^2$ and $3q^2$ represent the same natural number, which could be expressed as a product of prime natural numbers in two different ways, one employing an even number and the other an odd number of prime factors. Since this contradicts the fundamental theorem of arithmetic, $n$ is not rational. This type of argument applies to $n^2 = b$ for every $b$ which is the product of an odd number of prime factors.

3. (a) $0.\overline{8989}\cdots$; (b) $0.\overline{466557}$; (c) $0.\overline{587531498422}$; (d) $0.8989989998\cdots$; (e) $0.1111\cdots$

5. (a) 2, 3.1, 3.14, 3.146, 3.1463, $\cdots$
   (b) 1, 2.38, 2.4393, 2.449048, 2.44953582, $\cdots$

7. (a) Yes, except for the reciprocal of zero, which is not defined. The reciprocal of $m/n$ is $n/m$, $m \neq 0$.
   (b) If $b$ is the reciprocal of $c$ then $b \cdot c = 1$ which is rational. If $b$ is rational, then $c$, the reciprocal of $b$, is rational by part (a), so an irrational number does not have a rational reciprocal.

9. Yes, if the rational number is zero.

## Exercise 10.1, p. 189

1. (a) 0.06375; (b) 2.137; (c) $0.0566\cdots$; (d) 0.000035

3. (a) $\dfrac{51}{800}$; (b) $\dfrac{2137}{1000}$; (c) $\dfrac{17}{300}$; (d) $\dfrac{7}{200,000}$

5. (a) $0.225 \times 40 = 9$; (b) $n/100 \times 600 = 150$, so $n = 25$; 25%

7. (a) 3,007; (b) 194.0; (c) 0.009140

9. (a) 1.732; (b) 12.35; (c) 35.50; (d) $0.04163\cdots = 4.163\cdots \times 10^{-2}$ leaves doubt as to whether 0.04163 or 0.04164 is the approximation, so a four-digit approximation is not possible; (e) 107.5

## Exercise 10.2, p. 200

1. (a) $90.1570 < 43.76 + 19.034 + 27.369 < 90.1690$
   (b) $245,155 < 1.004 \times 10^5 + 8.38 \times 10^4 + 6.106 \times 10^4 < 245,365$
   (c) $15.695 < 3.45 + 2.17 + 4.63 + 2.38 + 3.09 < 15.745$

3. (a) $55.87\cdots < 41.07 \div 0.73 < 56.65\cdots$
   (b) $8.424\cdots < 9.689 \div 1.1 < 9.228\cdots$
   (c) $0.01014\cdots < 1.01 \div 99.04 < 0.01024\cdots$

5. (a) 0.318; (b) 0.000; (c) 1; (d) $100 \times 10^1$

7. (a) $k = 3$, so $n = 0$. Accuracy of 43.76 is $10^{-2}$, so $m = {}^-2$. Then $43.76 +$

19.034 + 27.369 = 90.163 is accurate to place value $10^{-2+0+1} = 10^{-1}$, so that 90.2 is a valid approximation.
(b) $10^{2+0+1} = 10^3$ place; $2.45 \times 10^5$
(c) $10^{-2+0+1} = 10^{-1}$ place; round off 15.72 to 15.7.

9. (a) $90.16 \pm 0.01$; (b) $245{,}260 \pm 105$; (c) $15.72 \pm 0.025$

*Exercise 10.3, p. 206*

3. (a) $(H, h), (H, t), (T, h), (T, t)$; (b) $\frac{1}{4}; \frac{1}{4}; \frac{2}{4} = \frac{1}{2}$;

(c) $\frac{1}{4} \times 20 = 5; 5; 10$

5. Record the two numbers showing on the two dice as ordered pairs with the number on one particular die always listed first. This procedure produces the set $\{(x, y) \mid x \in N_6, y \in N_6\}$ or $N_6 \times N_6$.

7. (a) $\frac{3}{16}$; (b) $\frac{5}{16}$; (c) $\frac{8}{16} = \frac{1}{2}$

9. (a) $\frac{3 \times 2}{6 \times 6} = \frac{1}{6}$; (b) $\frac{2 \times 1}{6 \times 6} = \frac{1}{18}$; (c) $\frac{1 \times 3}{6 \times 6} = \frac{1}{12}$;

(d) $\frac{3 \times 1 + 2 \times 2}{6 \times 6} = \frac{7}{36}$; (e) $\frac{2 \times 3 + 1 \times 1}{6 \times 6} = \frac{7}{36}$

*Exercise 10.4, p. 216*

1. (a) $1! = 1 \cdot 0! = 1 \cdot 1 = 1$; (b) $2! = 2 \cdot 1! = 2 \cdot 1 = 2$; (c) 6; (d) 24; (e) 5,040

3. $\binom{10}{3} = \frac{10!}{3!\,7!} = \frac{10 \cdot 9 \cdot 8 \cdot 7!}{3 \cdot 2 \cdot 7!} = 120$

5. $\frac{6!}{4!\,1!\,1!} \div 6! = \frac{1}{4!} = \frac{1}{24}$

7. (a) $\dfrac{10!}{3!\,7!}\left(\dfrac{1}{2}\right)^3\left(\dfrac{1}{2}\right)^7 = \dfrac{15}{2^7} = \dfrac{15}{128}$; also by $\binom{10}{3} \div 2^{10}$ (b) $\binom{10}{3}\left(\dfrac{1}{2}\right)^{10} +$

$\binom{10}{2}\left(\dfrac{1}{2}\right)^{10} + \binom{10}{1}\left(\dfrac{1}{2}\right)^{10} + \binom{10}{0}\left(\dfrac{1}{2}\right)^{10} = (120 + 45 + 10 + 1)\left(\dfrac{1}{2}\right)^{10} = \dfrac{11}{64}$

9. (a) $\binom{52}{5} = \dfrac{52 \cdot 51 \cdot 50 \cdot 49 \cdot 48 \cdot 47!}{5 \cdot 4 \cdot 3 \cdot 2 \cdot 1 \cdot 47!} = 5{,}197{,}920$; (b) $\binom{16}{5} = 4{,}368$;

(c) $4{,}368 \div 5{,}197{,}920 = \dfrac{1}{1{,}190}$

11. Both expressions represent the number one.

# *Bibliography*

Courant, R., and H. Robbins, *What Is Mathematics?*. New York: Oxford University Press, Inc., 1941.

Eves, Howard, *An Introduction to the History of Mathematics*, rev. ed. New York: Holt, Rinehart & Winston, Inc., 1964.

Hoel, Paul G., *Elementary Statistics*. New York: John Wiley & Sons, Inc., 1960.

Jones, B. W., *Elementary Concepts of Mathematics*, 2nd ed. New York: The Macmillan Company, New York, 1963.

Moise, Edwin E., *Elementary Geometry From An Advanced Standpoint*. Reading, Mass.: Addison-Wesley Publishing Co., Inc., 1963.

Peterson, John A., and Joseph Hashisaki, *Theory of Arithmetic*, 2nd ed. New York: John Wiley & Sons, Inc., 1967.

Richardson, Moses, *Fundamentals of Mathematics*, 3rd ed. New York: The Macmillan Company, 1966.

Stabler, E. R., *An Introduction to Mathematical Thought*. Reading, Mass.: Addison-Wesley Publishing Co., Inc., 1953.

# *Index*

Abacus, 72
Addition
  approximations, 190
  base six, 75
  cardinal numbers, 33, 34
  clock, 71
  distances, 85
  integers, 131
  number line, 85
  rationals, 144
  reals, 182
Algorithm, 48
  addition, 76
  common fractions, 155
  division, 116, 120, 168
  factoring, 154
  greatest common divisor, 150
  multiplication, 96, 98
  round off, 191
  subtraction, 112
And (logic), 4, 5, 6
Angle, 103
  proper, 103
  right, 104
  straight, 103
  zero, 103
Approximation, 187, 189
  addition, 190
  subtraction, 194
  multiplication, 197, 198
  division, 200

Area, 105
Aristotle, 3
Associative, 57, 70
Axiom, 18

Base, 43
Between, 82

Cantor, 180
Cardinal, 26, 31
Cartesian
  coordinate, 102
  product, 89
Carrying, 76
Cauchy, 180
Closed, 56, 69
Combination, 216
Commutative, 57, 69
Complement, 22
Conclusion (logic), 8
Contrapositive, 11
Converge, 180
Converse, 10, 11
Coordinate, 86
Corollary, 58, 142
Correspondence, 26
Counting, 29

Decimal, 42, 43
  repeating, 171
  terminating, 170

## 242   Index

Denominator, 147
Descartes, 190
Digits, 41, 42
  decimal, 41
  significant, 195
Disjunction, 5
Distance, 80, 82
Distributive, 61, 94
Dividend, 116, 147
Divides, 150
Division, 116
  algorithm, 116, 120
  base six, 125
  by zero, 116
  integers, 139
Divisor, 116, 147
  greatest common, 151

Elements, 15, 18
End point, 83
Equals, 16
Equation, 130
Equivalent
  equations, 130
  sets, 26
  statements, 8
Euclid, 117
Events, 201
  composite, 207, 209
  compound, 207
  dependent, 207
  equally likely, 202
  independent, 207
  mutually exclusive, 203
  random, 204
  simple, 207
Extension (system), 130

Factor, 151, 154
Factorial, 214
False, 3
Fermat, 200
Field, 184
Figure
  geometric, 87
  plane, 103

Fraction
  common, 147
  standard, 149

Geometry, 80
  analytic, 102
Graph, 102
Greater than
  cardinal number, 32
  distance, 84
  number, 160

Hypothesis, 8

Identity, 60, 92, 93
If and only if, 11, 12
Implication, 8
Implies, 4, 5
Induction, 54, 55
Integer, 131
  negative, 131
  positive, 131
Interior
  angle, 103
  region, 106
Intersection, 19
Inverse
  element, 137
  operation, 110
Irrational, 179

Laplace, 201
Lemma, 142
Line
  geometric, 102
  number, 79, 84
Logic, 2, 3

Multiplication, 60
  algorithm, 96
  base four, 99
  integers, 132
  lattice method, 98
  number line, 101
  rationals, 142

## Index

Negation, 6
Negative
  integer, 131
  number, 159
  of a number, 137
  rational, 157
Notation
  expanded, 45, 96
  factorial, 214
  scientific, 188
  set builder, 17
Null set, 22
Number, 25
  algebraic, 179
  cardinal, 26, 31
  composite, 153
  counting, 50
  irrational, 179
  natural, 28
  negative, 159
  ordinal, 29
  positive, 159
  prime, 153
  rational, 141
  real, 179, 181
  transcendental, 179
  whole, 28, 56
Number line, 79, 84
Number theory, 149
Numeral, 27, 38, 42
  decimal, 41
  Egyptian, 38
  fraction, 147
  Greek, 41
  Hindu-Arabic, 41
  Roman, 40
Numerator, 147

Open sentence, 3, 129
Operation, 66
  binary, 66, 67
  inverse, 110, 116, 122
  modular, 145
  ternary, 71
  unary, 71
  well defined, 66
Opposite, 10, 11, 137

Opposite converse, 11
Or (logic), 4, 5
Order, 27, 158
Ordinal, 29, 50

Partition, 115
Pascal, 200, 216
Peano, 53
Per cent, 186
Permutation, 213
Place value, 42, 47
Plane, 108
Point
  decimal, 169
  geometric, 80
Prime
  number, 153
  relatively, 151
Probability, 205
  binomial, 211
Product
  cardinal number, 91
  partial, 97
  reals, 182
  set, 89, 90

Quadrilateral, 104
Quotient, 116, 147

Ratio, 147
Rational numbers, 141
  equality, 145
Ray, 83
Reciprocal, 140
Rectangle, 104
Region, 105
Regrouping, 76
Remainder, 117

Segment, 83
Sentence, 2
  open, 3, 129
Sequence, 66
  Cauchy, 180
  fundamental, 180
Set, 2
  disjoint, 22
  equivalence, 26

finite, 28, 31
infinite, 28, 31
notation, 16, 17
null, 22
product, 89, 90
solution, 129
subtraction, 20
universal, 22
Side
  angle, 103
  quadrilateral, 104
Solution, 129
Space, 80
Square, 104
Statement, 2, 4
Straight edge, 82
Subset, 20
  proper, 20
Subtraction
  approximations, 194
  base six, 122
  cardinal number, 35
  integers, 138
  inverse to addition, 110
  number line, 86
  sets, 20, 35
Successor, 53
Sum
  cardinal number, 34
  distance, 85
  partial, 77
  reals, 182

Surface, 102
System
  binary, 43
  integers, 131
  mathematical, 56, 69
  numeration, 42
  place value, 42
  rational, 142
  reals, 183
  whole numbers, 56, 91, 93

Tables:
  addition
    base four, 99
    base six, 75
    base seven, 100
  multiplication
    base four, 99
    base seven, 100
  truth, 5
Theorem, 142
True, 2

Union, 19
Universal set, 22

Valid, 13
Venn diagrams, 22
Vertex (vertices), 103, 104

Zero, 27